D1238781

Pest Management Programs for Deciduous Tree Fruits and Nuts

Pest Management Programs for Deciduous Tree Fruits and Nuts

Edited by

David J. Boethel

Louisiana Agricultural Experiment Station
Pecan Research and Extension Station
Shreveport, Louisiana

and

Raymond D. Eikenbary

Oklahoma State University
Stillwater, Oklahoma

PLENUM PRESS · NEW YORK AND LONDON

Library of Congress Cataloging in Publication Data

Main entry under title:

Pest management programs for deciduous fruits and nuts.

Includes bibliographical references and index.
1. Fruit – Diseases and pests – Congresses. 2. Nuts – Diseases and pests – Congresses. 3. Insect control – Congresses. 4. Pest control, Integrated – Congresses. 5. Mites – Control – Congresses. I. Boethel, David J.
SB608.F8P46 634'.104'9652 79-12616
ISBN 0-306-40178-9

Based on papers presented at the 1977 National Meeting of the Entomological Society of America and additional invited papers

A Division of Plenum Publishing Corporation
227 West 17th Street, New York, N.Y. 10011

Printed in the United States of America

Contributors

D. Asquith	Fruit Research Laboratory, Pennsylvania State University, Biglerville, Pennsylvania, 17307.
M. M. Barnes	Department of Entomology, University of California, Riverside, California, 92502.
D. J. Boethel	Louisiana Agricultural Experiment Station, Louisiana State University Pecan Research and Extension Station, Shreveport, Louisiana, 71105.
B. A. Croft	Department of Entomology, Michigan State University, East Lansing, Michigan, 48824.
R. K. Curtis	Almond Board of California, Sacramento, California, 95813.
C. S. Davis	California Cooperative Extension Service, University of California, Berkeley, California, 94720.
R. D. Eikenbary	Department of Entomology, Oklahoma State University, Stillwater, Oklahoma, 74074.
H. C. Ellis	Georgia Cooperative Extension Service, University of Georgia Coastal Plain Experiment Station, Tifton, Georgia, 31794.
L. A. Hull	Fruit Research Laboratory, Pennsylvania State University, Biglerville, Pennsylvania, 17307.
M. L. Marino	Department of Botany and Plant Pathology, Michigan State University, East Lansing, Michigan, 48824.
J. R. McVay	Alabama Cooperative Extension Service, Auburn University, Auburn, Alabama, 36830.
D. J. Miller	Department of Entomology, Michigan State University, East Lansing, Michigan, 48824.

H. Riedl — Department of Entomology, University of California, Berkeley, California, 94720.

S. M. Welch — Department of Entomology, Kansas State University, Manhattan, Kansas, 66506.

P. H. Westigard — Southern Oregon Experiment Station, Oregon State University, Medford, Oregon, 97502.

Preface

Pest Management Programs for Deciduous Tree Fruits and Nuts attempts to present the current status of pest management programs in orchard ecosystems. The book is a collection of papers from a symposium convened on the subject for the 1977 National Meeting of the Entomological Society of America and invitational papers on commodities not covered during the symposium.

In recent years, books have appeared on "integrated pest management (IPM)"; however, most of these have concentrated on field crop IPM with an occasional chapter on fruits. No publication presently exists which brings together information on the pest management programs currently being conducted on the major nut crops, almonds, pecans and walnuts.

Because it is the first treatment for almonds and walnuts, the authors of these chapters have attempted not only to present the current IPM technology but the historical data which led to the contemporary programs.

Two chapters appear on pecan IPM. The first concentrates on the development of a management program for the pecan weevil, the key arthropod pest of pecans, while the second discusses the implementation of pilot pecan IPM programs in two southeastern states. The latter chapter illustrates that even with a limited data bank, the pesticide load in pecan orchards can be reduced by the adoption of the IPM approach to pest control.

Although deciduous tree fruits have received more recent exposure, the dynamic nature of IPM on apples and pears certainly warrants inclusion in this work. Today, many of the most sophisticated IPM programs are those practiced in these ecosystems. Two chapters are devoted to case histories of the development of pest management systems in Oregon pear orchards and Pennsylvania apple orchards. The final chapter presents an elaborate discussion of the extension pest management delivery and biological monitoring system research underway in Michigan. Even though this chapter uses apple IPM as an example, the information certainly is worthy of consideration by all personnel involved in IPM regardless of the pest or crop situation.

Our objective when organizing the symposium was to assemble workers on the various orchard crops in order to share ideas on IPM. This book is an attempt to provide something tangible to accomplish the objective. We sincerely thank the contributors for their efforts in this endeavor. The symposium participants have expanded their original presentations, and these manuscripts along with the invitational papers constitute the most comprehensive reference material available on orchard IPM. The book should serve as a reference text for graduate and undergraduate courses in integrated pest management and should be a valuable addition to the libraries of researchers, extension personnel, industry representatives, and fieldmen involved in crop protection.

January, 1979

D.J.B.
R.D.E.

Contents

Chapter III

Status of Pest Management Programs for the Pecan Weevil

D. J. Boethel and R. D. Eikenbary

Chapter IV

Extension Approaches to Pecan Pest Management in Alabama and Georgia

J. R. McVay and H. C. Ellis

Chapter V

Integrated Pest Management of Insects and Mites of Pear

P. H. Westigard

Chapter VI

Integrated Pest Management Systems in Pennsylvania Apple Orchards

D. Asquith and L. A. Hull

Chapter VII

Developments in Computer-Based IPM Extension Delivery and Biological Monitoring System Design

B. A. Croft, S. M. Welch, D. J. Miller, and M. L. Marino

PRESENT DEVELOPMENT OF ARTHROPOD PEST MANAGEMENT IN ALMOND ORCHARDS OF CALIFORNIA

Martin M. Barnes
Department of Entomology
University of California
Riverside, California

Robert K. Curtis
Almond Board of California
Sacramento, California

INTRODUCTION

The cultivated almond, *Prunus amygdalus* Batsch, apparently originated from selections of *P. communis* Fritsch which occurs in southeastern Russia, Afghanistan and Iran, and its subsequent cultivation in the Mediterranean area pre-dates recorded history. Although introduced into California from Spain via Mexico when the missions were established, these trees died as the missions declined. In 1840, trees were imported from Europe and planted in the New England and mid-Atlantic states but without success as the climate is too severe. In 1843, almond trees were brought from the east coast of the U.S. and planted in California, marking the beginning of present industry (Kester and Asay 1975; Meith *et al.* 1977).

At present, virtually all of the commercially produced almonds in the United States are grown under irrigation in California, where the main producing areas are in the confluent valleys of the San Joaquin and Sacramento Rivers, hereinafter referred to as the "central valley" (Fig. 1). There are 341,000 acres planted, of which about 80% are of bearing age (Anon. 1978b), producing 313 million pounds of nutmeats in 1977 (Anon. 1978a).

THE ALMOND ORCHARD ECOSYSTEM

Although the almond tree is almost as hardy as the peach, climatic requirements for good crop production are rather exacting. The almond has a low chilling requirement and its period of bloom (February 10 to March 1 in

Figure 1. Distribution of almond acreage in the central valley of California. Each dot represents 1000 acres of bearing trees (California Crop and Livestock Reporting Service, 1977).

California valleys) makes the flowers susceptible to frost injury, probably the most important factor in defining crop regions. A Mediterranean-type climate with a rainy, mild winter followed by a warm, rainless spring and summer provides ideal conditions (Kester and Asay 1975). It is well suited to California's hot interior valleys where rainfall is abundant during mild winters, and in spring, low rainfall and humidity suppress infection by disease organisms. Nevertheless, several plant pathogens affect almond trees of the area, and their symptomatology, field biology and control are discussed by Moller *et al.* (1976). Rain is rare in summer in the central valley, favoring maturation of the nuts and hull dehiscence, and rather infrequent in early fall, providing favorable conditions for harvest. Almond trees continue to grow in the fall if there is adequate moisture and warm weather. Fruit bud differentiation initiates in summer and continues through fall and winter (Wood 1947). Therefore mite control is important to minimize leaf injury and defoliation before and after harvest.

The almond has the three parts which typify a drupe fruit. The exocarp is pubescent, and the mesocarp or hull is fleshy and dehisces as it matures and becomes dry and leathery as it opens exposing the nut. The endocarp or shell varies from hard to thin and paperlike. The seed or nutmeat contains the embryo surrounded by seedcoats (Kester and Asay 1975).

The principal variety in California is the 'Nonpareil,' a papershell almond very susceptible after hull dehiscence (hullsplit) to entrance by lepidopterous larvae. All major varieties are self incompatible and adequate cross-pollination by a cross-compatible variety is essential. Three varieties are usually planted so that bloom periods may overlap. A common pattern is two rows of 'Nonpareil' alternated first with one row of an early blooming variety (*e.g.* 'Ne Plus Ultra') and then with a late blooming variety (*e.g.* 'Mission'). Pollination is carried out predominantly by honey bees, averaging 2½ hives per acre. These are rented from beekeepers and installed in orchards at the onset of bloom (Thorp and Mussen 1977). Guidelines for the protection of honey bees from pesticides are presented by Atkins *et al.* (1975 and 1977).

Almonds are deep rooted and grow and produce best on deep, well drained soils of light to medium texture. In most new plantings on good soils, about 75 trees are planted per acre. Irrigation is the most important cultural factor of almond production in California and non-irrigated orchards, once extensive, are now rare. During a typical season in the hot central valley, mature almond trees will use 35 inches of water per acre (Meith *et al.* 1977). This is applied by flood or sprinkler irrigation for the most part; however, drip irrigation systems requiring less water are coming into use.

A combined nontillage and strip weed control system is extensively used with conventional irrigation practice. The components are a volunteer winter cover crop between rows, which is mowed frequently, and a herbicide treated weed-free strip down the tree row (Meith and Brown 1975). In addition to horticultural advantages, this practice reduces dust in the orchard, which otherwise tends to induce spider mite problems. Volunteer cover crops are common and annual grasses are preferred, as equipment for pest control operations can enter the orchard sooner after a rain or irrigation than if a broadleaf cover or no cover crop is grown (Meith *et al.* 1977).

Hullsplit initiates on the 'Nonpareil' variety from early July to early August, depending on latitude and soil conditions, and the period of hullsplit and hull drying prior to harvest extends over four to five weeks. In preparation for harvest pickup, water is withdrawn. The floor of the orchard may then be levelled and freed of vegetation. After harvest, irrigation is resumed.

'Nonpareil' harvest begins in mid-August and harvest is predominantly mechanical, using tree shakers, sweepers, and pickup machines, followed by delivery to the huller. In this and in other respects, almonds are an industrialized farm commodity. Almond hulls contain 25% sugar and are valuable as components of livestock feed (Weir 1951). A recent trend is to add shells to the ration; hence, insecticide residues must fall within tolerance levels.

Most pollinizer varieties dehisce and mature after the 'Nonpareil' and their harvest takes place in September and October. How this complicates insect pest management practice will be related below.

THE ARTHROPOD PEST COMPLEX

In comparison with other deciduous tree crops, the almond orchard has relatively few major arthropod pests, although these may cause severe losses.

Key Pests

Navel Orangeworm. The most important pest is the pyralid moth, *Amyelois transitella* (Walker), commonly known as the navel orangeworm. Losses in 1977 are estimated at $24,000,000 (Anon. 1978a). This species first came to prominence in the U.S.A. infesting damaged navel oranges in Arizona in 1921, but it is not a pest of oranges (Wade 1961). It is primarily a scavenger and the larvae are found in a wide range of fruits, nuts and legume pods which are dried and mummified, decayed, or damaged by other insects. Though the scavenging habit predominates, the navel orangeworm infests and develops well on sound, new crop almonds, walnuts, and pistachios after hullsplit. Prior to hullsplit, its populations develop in almond orchards in mummy fruits which remain on the tree after harvest and persist through the following year.

Overwintering takes place predominantly in the larval state in the mummy nuts. Some feeding and development take place during warmer periods of winter as there is no diapausing stage. Moths of the overwintered brood usually begin to emerge in late March and April, but the principal period of emergence and egg laying occurs in May and is completed by mid-June, depending on season and latitude (Caltagirone *et al.* 1968; Rice 1976). The moth may disperse over considerable distances; female movement has been observed extending to 300 meters in two nights. Infestation levels in pistachios, previously non-bearing and adjacent to infested almonds, were detected in a gradient up to one mile (Andrews 1978). Overwintered brood moths lay their eggs on mummy nuts and the first brood develops almost exclusively in these. The moths of the first brood, which begin to emerge in late June, also oviposit on mummy nuts. However, with the initiation of hullsplit of the 'Nonpareil' variety, oviposition of this brood turns to the new crop.

After hullsplit is complete, about half the eggs of the second generation are laid on the outside of the hull, one-third on the shell and the remainder on the interior of the hull. Furthermore, as the infestation develops, the moths prefer to oviposit on previously infested nuts. Moths of the second generation, developing from eggs laid on the new crop, begin to emerge 30 days after oviposition with 50% emergence completed in 43 days (Caltagirone *et al.* 1968; Curtis and Barnes 1977). Consequently, second generation moths developing on the new 'Nonpareil' crop emerge in time to reinfest this crop, emphasizing early harvest as a key management procedure. Also, these second generation moths are at peak emergence and oviposition before the later harvest of the pollinizer varieties is begun. Thus, later varieties are subject to heavier infestation pressure. In 1977, the 'Merced' and the 'Thompson' varieties, which are harvested after 'Nonpareil,' averaged 11% and 13% navel orangeworm damage, respectively, as compared with 6% for 'Nonpareil' (Anon. 1978a).

Varietal susceptibility is closely correlated with shell characteristics as well as later harvest period. The most susceptible variety, given equal population pressure, is the 'Nonpareil,' which often has shell perforations. When subject to equivalent infestation pressure, the papershell 'Nonpareil' variety suffered an average 11.3% infested kernels while the hardshell 'Peerless' variety, which is harvested at the same time, suffered 4.3% (Crane and Summers 1971). The 'Mission' is a late harvested, hardshell variety which is highly resistant to navel orangeworm, but it does not have the same marketing characteristics as 'Nonpareil.'

Peach twig borer. This gelechiid moth, *Anarsia lineatella* Zell., was thoroughly studied by Bailey (1948). It is reported throughout the United States and also occurs in Europe and the Near East. It may develop dense populations in untreated orchards and frequently causes severe damage to nutmeats of papershell and softshell varieties (Summers and Price 1959; Summers 1962). Prior to the development of organic insecticides, statewide damage ranged from 10% to 29% over a 12-year period (Bailey 1948). The species overwinters as immature larvae, primarily second instar, in hibernacula on the tree. The overwintered larvae emerge during bloom, feed on buds and shoots, and emerge as moths in April and May. The majority of first brood eggs hatch in May and the larvae mine the new shoots and occasionally infest green nuts. Second brood larvae appear in late June or early July and mine shoots as well as attack the ripening 'Nonpareil' fruit. The fruit is first entered just prior to hullsplit through the stem end where the hull has loosened from the spur. After hullsplit, larvae are found feeding on the green hulls and ripening nutmeats but new infestation declines as these dry. Third brood larvae appear in September and October on later maturing varieties. Some of these develop a partial fourth generation in November (Rice and Jones 1975) which overwinters, while the rest enter hibernacula directly.

Mites. Six species of tetranychid mites infest almond trees in the central valley - the European red mite, *Panonychus ulmi* (Koch), the citrus red mite, *P. citri* (McGregor), the brown almond mite, *Bryobia rubrioculus* (Scheuten), and three species of webspinning mites, the twospotted spider mite, *Tetranychus urticae* Koch, the Pacific spider mite, *T. pacificus* McGregor, and the strawberry spider mite, *T. turkestani* Ugarov & Nikolski. The brown mite and European red mite may reach pest status in the central and northern regions of the central valley (Summers 1962; Hoy *et al.* 1978) but are seldom encountered in the southern part. Summers (1962) reported that the European red mite apparently was "adapting" to almond and appeared more frequently as a pest. This might also have been related to developing patterns of insecticide use at that time. Citrus plantings are largely confined to the southeastern part of the central valley, and females of the citrus red mite "balloon" from citrus to infest almonds in spring. Damaging populations of this species develop in many orchards. This species does not persist on almond during winter and annually invades almond orchards from citrus (Barnes and Andrews 1978).

All three webspinning species are present in the southern valley. In that area they are important key pests, developing high populations annually even when not disturbed by use of insecticides. In the northern part of the valley, twospotted spider mite and Pacific spider mite are present. Severe infestations occur but are not so frequent there as in the south (Summers 1962).

Significant populations of an eriophyid, the peach silver mite, *Aculus cornutus* (Banks), develop on almond foliage in the northern central valley but are not believed to be damaging (Hoy *et al.* 1978).

Minor Pests

Several other species of insects may cause minor losses. The fruittree leafroller, *Archips argyrospilus* (Walker), and the western tent caterpillar, *Malacosoma californicum* (Packard), are rather infrequently encountered (Summers 1962). The boxelder bug, *Leptocoris trivittatus* (Say), may cause drop of small fruit, and kernel spotting and gumming of older nuts in areas where boxelder, oak and elderberry abound, invading orchards soon after bloom. As well, the leaffooted plant bug, *Leptoglossus clypealis* Heidemann, and the consperse stink bug, *Euschistus conspersus* Uhler, occasionally invade orchards and cause kernel spotting and gumming.

Almonds are not generally troubled with scale insects, but local outbreaks of three common species may occur: the San Jose scale, *Quadraspidiotus perniciosus* (Comstock), the olive scale, *Parlatoria oleae* (Colvee), and the European fruit lecanium, *Lecanium corni* Bouche (Summers 1962).

PRESENT STATUS OF COMPONENTS OF THE INSECT PEST MANAGEMENT SYSTEM

Biological Monitoring

Oviposition traps are a useful method for monitoring activity of the female navel orangeworm moths. These traps, which are baited with the laboratory larval rearing medium for the species, accurately reflect seasonal egg laying patterns in orchards until hullsplit of the new crop begins (Fig. 2) (Rice 1976; Rice *et al.* 1976). After hullsplit, the attractiveness of the new crop greatly

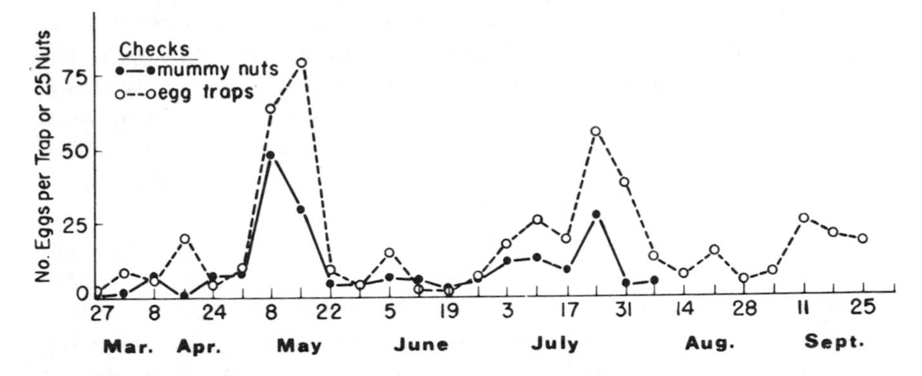

Figure 2. Comparison of egg deposition trends by navel orangeworm on almond mummies and egg traps in an untreated orchard during 1974 (Rice 1976).

diminishes oviposition on the egg traps. They are commercially available and may be used to indicate the time for treatment against the first generation in spring. Information is being developed on heat summation in spring in relation to emergence and activity of moths and egg hatch (Barnes 1978; Rice 1978). Research is currently underway toward chemical identification of the sex pheromone produced by the female moth (Tumlinson *et al.* 1978).

The major component of the female sex pheromone of the peach twig borer has been identified (Roelofs *et al.* 1975) and field studies have demonstrated that a 5:1 mixture of *trans*-5-decenyl acetate and *trans*-5-decen-1-o1 was best for attracting male moths (Rice and Jones 1975). Sex pheromone traps are useful in monitoring the appearance of moths of the overwintering generation and may be used for timing treatments against first brood larvae.

Significance of Infestation Levels

Navel orangeworm larvae feed on the kernels causing damage ranging from a small pinhole entrance to virtual destruction. They produce a large amount of webbing and frass within infested nuts. A proportion of damaged nuts are removed from the crop by air separation in the pickup and hulling operations and are not recorded in samples taken after hulling. Eggs present on the hulls and nuts at the time of tree shaking may hatch resulting in increased infestation. The nuts may not be swept from the orchard for several days, nor hulled for a period of time, because present huller capacity cannot keep pace with harvested tonnage. At present, a decision on whether to take action for navel orangeworm suppression depends on (a) infestation history of the previous season, (b) presence of a significant population of mummy nuts in the orchard, and (c) because of the wide dispersal habits of the moth, proximity of the orchard to other sources of infestation.

Peach twig borer larvae cause shallow channels and surface etching on nutmeats. This distinguishes their damage to nutmeats from that caused by navel orangeworm larvae, which bore into the nutmeat. Measures against peach twig borer are especially important on young trees. Populations causing damage to nutmeats are very common in bearing orchards and insecticides are applied during the dormant period or against the first brood on a large proportion of the acreage. The relationship between pheromone trap moth catches in spring and subsequent levels of infestation, *e.g.* twig mining and nutmeat damage, is not known.

Severe mite infestations develop in July and August in the southern part of the valley and if not controlled, result in partial defoliation. There is no effect on crop size or quality the first year of such infestation. However, two well-replicated experiments have shown that in the following year the crop was reduced an average of 16%, and both terminal and trunk girth growth were decreased (Barnes and Andrews 1978). Unless mites are under satisfactory control, withdrawing irrigation water and drying orchards in preparation of the orchard floor for harvest will intensify defoliation related to mite damage. Although predacious insects are present and affect some regulation of mite populations in the central and northern parts of the valley (Hoy *et al.* 1978), information is needed as to what ratios of prey to predator would indicate a favorable outcome. Nor is information available at present which provides for decision making based on the relationship of the mean number of mite-days per leaf to which the tree has been subjected and effects on tree growth and productivity. However, Andrews and La Pre (1979) have developed preliminary information on the effects of various levels of Pacific mite-days per leaf and rate of photosynthesis and transpiration of almond leaves. This approach has a clear potential for development in the analysis of threshold effects of mites on almond as well as on other plants, as the equipment which was used is portable and readily employed under field conditions.

Cultural Interventions

Two cultural procedures in the orchard play an important role in management of the navel orangeworm, (a) orchard sanitation, and (b) early harvest (followed as soon as possible by fumigation).

Since the navel orangeworm maintains its populations on mummy almonds on the tree through much of the year, a concerted effort for removal of mummies (and subsequent destruction by discing or flailing) in winter can significantly suppress damage. Not only is there direct mortality from proper sanitation practice, but the number of mummies on the tree available for colonization by the overwintering brood emerging in spring is reduced. The cultural approach to management of this insect has been explored by C. E.

Curtis (1976, 1978). Mummy almonds can be shaken from trees in winter by tree shakers and hand poling, and removal is best on days when heavy fogs soak the dead abscission layer by which the nut is attached. Results from sanitation efforts in a 40-acre block which was not isolated from other orchards reduced infestation on 'Nonpareil' by approximately one-half. In an isolated orchard, a severe infestation was reduced, in successive years, from 37% to 5% (Curtis 1976).

Early harvest followed by fumigation is critical to any successful program of management of the navel orangeworm. As previously noted, there is an explosive increase in the moth population in the period beginning 30 days after the initiation of hullsplit. At this time, emergence of second generation moths begins, those developing from oviposition at hullsplit. Survival and biotic potential of this generation is high, as an ample food supply is available in the new crop. These moths lay the eggs of the third generation from which infestation builds rapidly. It may be diminished, however, by early harvest and ensuing damage reduced by rapid delivery to fumigation. It is important to note that the benefits of suppression by cultural or chemical means will be lost if the nuts remain on the trees for an extended period of time after they are ready for harvest.

The "Ballico project," a large scale effort covering 12 square miles and 60 growers, was conducted by the USDA, SEA, to demonstrate the combined effectiveness of orchard sanitation and early and rapid harvest to suppress navel orangeworm damage in almond orchards. Mummy fruits on navel orangeworm hosts such as almond, walnut, peach, and plum were removed during December and January by trunk-shakers and hand poling. Early harvest by growers was encouraged in the test area. An area of nine square miles involving 63 growers was observed as a check. Results in 1977, the third year of the project, showed a mean infestation of the 'Nonpareil' variety in the "treated" area of 4.4% as compared with 8.8% in the check area.

An important criterion for evaluating performance of new varieties is shell seal and hardness, and work has been done to this end (Kester and Asay 1975; Kester 1978). Attention has been given to laboratory measurements of the seal of almond shells, which correlate with resistance of the whole nut to penetration by navel orangeworm (Soderstrom 1977). This method of determining resistance will be useful in almond breeding programs.

Because of the heavier navel orangeworm infestation found on papershell and softshell pollinizer varieties maturing later than 'Nonpareil,' resulting from the increase in the population of this species developing on 'Nonpareil' after the earlier hullsplit of this variety, it is apparent that progress in management of this pest can be made by the development of cross-compatible varieties which bloom together and are harvested together. Such a system would require that the varieties be sufficiently similar in kernel characteristic that they can be marketed as a single product.

Biological Interventions

Searches have been made for parasites of the navel orangeworm in Mexico and Texas, and the polyembryonic encyrtid, *Pentalitomastix plethoricus* Calt., was found (Caltagirone *et al.* 1964). This parasite is apparently a weak searcher, depending more on its polyembryonic characteristic for survival rather than on its searching ability. Introductions into California orchards were undertaken, but after more than ten years of releases, the wasp's ability to effectively limit navel orangeworm activity so that damage is reduced has not been demonstrated. Efforts made to mass rear and inundate orchards with this parasite did not yield suppression. Another polyembryonic parasite, *Paralitomastix pyralidis* (Ashm.), was imported and liberated against the peach twig borer in 1932 but was rarely collected in ensuing years (Bailey 1948).

Exploration is currently underway to locate other species which may attack the navel orangeworm (Legner 1978). Attention is also being given to importation and establishment of a *Stethorus* sp. to enhance mite predation in almond orchards (Hoy 1977).

Biological Control

No significant native parasitization or predation of navel orangeworm has been reported. When the mummy nut population has been greatly reduced by winter sanitation efforts, birds will remove a significant proportion of the remaining nuts in some areas.

Bailey (1948) lists a large number of parasites and several predators of the peach twig borer occurring in California and made an intensive study of their activity. In the hibernacula, nearly all the parasites of larvae found on almond were *Hyperteles lividus* (Ashm.), and this species was recorded as potentially the most effective. Among predators, the mite, *Pediculoides ventricosus* (Newp.), is listed as significant, also attacking larvae in hibernacula. In his study, Bailey (1948) concluded that although biological control was at times important, peach twig borer infestations were generally severe.

In the southern central valley where mites are key pests, the sixspotted thrips, *Scolothrips sexmaculatus* (Pergande), greatly reduces mite populations, but only after serious damage has been done (Barnes and Andrews 1978). In the north, biotic regulation by coccinellids, chrysopids and thrips is a more important factor (Hoy *et al.* 1978).

Chemical Interventions

Applications are made by concentrate or full coverage airblast sprayers and are generally in the range of 100 to 500 gallons per acre. The best timing for control of peach twig borer, scale insects, and eggs of European red mite and brown mite is during the dormant period, as spraying at that time provides the best coverage and is least likely to suppress beneficial species, *e.g.* mite predators in orchards of the northern central valley. An emulsive, dormant grade spray oil is used in combination with an organophosphate such as azinphosmethyl, diazinon, parathion, phosmet or carbophenothion. When using these treatments, honey bees should not be in the orchard. The hazard is great if the winter cover crop or annual weeds are in bloom. An alternate timing for peach twig borer, when weather conditions force omission of the dormant treatment, is in May, directed against moths of the overwintered generation and first generation larvae.

Chemical control of the navel orangeworm has been of assistance only in conjunction with a program of early and rapid harvest. Spray treatments have achieved reductions in infestation in the range of 50% to 85%, but this advantage is lost if harvest is delayed, exposing the new crop to heavy infestation by the larvae of the third generation. Two opportunities arise for insecticide treatment against navel orangeworm. The first is during the oviposition period of the overwintered brood moths. As noted previously, this is timed by egg traps which detect the cycle of oviposition in spring that principally falls in May. Considerable evidence exists that this timing also suppresses the peach twig borer adequately, although additional information is needed on how closely spring emergence cycles of the two species coincide. Azinphosmethyl is preferred at present for this treatment. Azinphosmethyl may also be applied in mid-June (60 days before harvest) but more evidence on the effectiveness of this timing is needed. When mite populations are present, azinphosmethyl may increase them (Hoy *et al.* 1978) and an acaricide, *e.g.* propargite or cyhexatin, should be included. Phosmet is also registered for spring use, but more evidence on effectiveness is needed. Use of carbaryl in spring should be avoided, as mite populations are so greatly intensified by early season use as to be virtually uncontrollable.

The second opportunity for partial suppression by insecticides is just prior to or at the initiation of hullsplit of 'Nonpareil'. Carbaryl and phosmet are in use at present for hullsplit treatments (28 or 30 days before harvest, respectively) and offer 50% suppression. This degree of control is the best that can be expected, since after hullsplit half the eggs are laid within the hull or on the shell. Spray coverage within the opened hull is fractional at initiation of hullsplit and later applications of currently registered insecticides have residue problems.

When carbaryl is used, an acaricide must be added; otherwise severe damage from mites will ensue. The later maturing pollinizers cannot be treated at the time of initiation of their hullsplit because the spray would drift onto adjacent unharvested 'Nonpareils,' which have a minimum 28- or 30-day interval to harvest for currently registered insecticides. Treatment may be made on these varieties in some circumstances after 'Nonpareils' are harvested.

As previously noted, foliage treatment for control of *Tetranychus* mites is required annually in substantial portions of the almond acreage. Timing is at the onset of significant infestations and usually falls in late June or early July. The acaricide treatment, currently consisting of propargite or cyhexatin, may often be combined with the hullsplit treatment for the navel orangeworm. Propargite and cyhexatin are selective and, from their spectrum of activity on other crops, it is assumed that when they are used alone for mite control they permit substantial survival of any mite predators which may be present.

Post-harvest Fumigation

Virtually the entire almond crop is fumigated after harvest to prevent further damage by navel orangeworm larvae. This operation may take place on the farm prior to hulling, in which case aluminum phosphide, which generates hydrogen phosphide as the active ingredient, may be used (Nelson *et al.* 1978). Fumigation prior to hulling is important and needs further exploitation, as fumigation after hulling may be delayed by several weeks because the crop cannot all be hulled at once. After hulling, methyl bromide or hydrogen phosphide fumigation is used upon delivery to the processing plant.

PRESENT STATUS OF INTEGRATED PEST MANAGEMENT

Six interventions, two cultural and four chemical, provide the present level of arthropod pest management achievable for almond orchards in California; (a) orchard sanitation in winter, consisting of removal of the great majority of mummy almonds infested with navel orangeworm by machine shaking or hand poling the trees followed by discing or flailing. Hand poling may also be undertaken immediately after shaking, recovering these nuts for the harvest; (b) a dormant spray for peach twig borer, scale insects and overwintering mite eggs; or if a dormant spray is not applied, a spring treatment for peach twig borer; (c) a May or hullsplit insecticide treatment for navel orangeworm; (d) an acaricide spray in summer, often combined with the hullsplit spray, followed by (e) early harvest and rapid delivery to (f) fumigation. From information available at present, sanitation or chemical treatment alone or in combination does not suffice for navel orangeworm control unless practiced in combination with early harvest soon followed by fumigation.

REFERENCES

Anonymous. 1978a. Almond Board of California statistics. Sacramento, CA.

Anonymous. 1978b. 1977 California Fruit and Nut Acreage. *Calif. Crop and Livestock Rept. Serv.*, Sacramento, CA.

Andrews, K.L. 1978. Investigations on the biology, ecology, and management of various tetranychid mites and of the navel orangeworm, *Paramyelois transitella* (Walker), on almonds in the southern San Joaquin Valley of California. Ph.D. Dissertation, University of California, Riverside.

Andrews, K. L., and L. La Pre. 1979. Effects of Pacific spider mite on physiological processes of almond foliage. *J. Econ. Entomol.* (in press).

Atkins, E. L., D. Kellum, and K. J. Neuman. 1975. Toxicity of pesticides to honey bees. *Div. Agr. Sci., Univ. of Calif. Leaflet* 2286. 4 pp.

Atkins, E. L., L. D. Anderson, D. Kellum, and K. J. Neuman. 1977. Protecting honey bees from pesticides. *Div. Agr. Sci., Univ. of Calif. Leaflet* 2883. 15 pp.

Bailey, S. F. 1948. The peach twig borer. *Calif. Agr. Exp. Sta. Bull.* 708. 56 pp.

Barnes, M. M. 1978. Unpublished report to the Almond Board of California, Sacramento, CA.

Barnes, M. M., and K. L. Andrews. 1978. Effects of spider mites on almond tree growth and productivity. *J. Econ. Entomol.* **71**:555-558.

Caltagirone, L. E., D. W. Meals, and K. P. Shea. 1968. Almond sticktights contribute to navel orangeworm infestations. *Calif. Agric.* **22(3)**:2-3.

Caltagirone, L. E., K. P. Shea, and G. L. Finney. 1964. Parasite to aid control of the navel orangeworm. *Calif. Agric.* **18(1)**:10-12.

Crane, P. S., and F. M. Summers. 1971. Relationship of navel orangeworm moths to hard shell and soft shell almonds. *Calif. Agric.* **25(1)**:8-9.

Curtis, C. E. 1976 and 1978. Orchard management and the Ballico/Famoso Project for navel orangeworm control. Unpublished reports to Almond Board of California, Sacramento, CA.

Curtis, R. K., and M. M. Barnes. 1977. Oviposition and development of the navel orangeworm in relation to almond maturation. *J. Econ. Entomol.* **70**:395-398.

Hoy, M. A. 1977. Unpublished report to Almond Board of California, Sacramento, CA.

Hoy, M. A., N. W. Ross, and D. Rough. 1978. Impact of navel orangeworm insecticides on mites in northern California almonds. *Calif. Agric.* **32(5)**: 10-12.

Kester, D. E. 1978. Unpublished report to Almond Board of California, Sacramento, CA.

Kester, D. E., and R. Asay. 1975. Almonds. pp. 387-419. *In* "Advances in Fruit Breeding." (J. Janick and J. N. Moore, eds.). Purdue University Press, West Lafayette, Ind. 623 pp.

Legner, E. F. 1978. Unpublished report to Almond Board of California, Sacramento, CA.

Meith, C., and T. Brown. 1975. Nontillage and strip weed control in almond orchards. *Div. Agr. Sci., Univ. of Calif. Leaflet* 2770. 9 pp.

Meith, C., W. C. Micke, and A. D. Rizzi. 1977. Almond production. *Div. Agr. Sci., Univ. of Calif. Leaflet* 2463. 20 pp.

Moller, W. J., W. C. Micke, and M. H. Gerdts. 1976. Almond Disease Guide. *Div. of Agr. Sci., Univ. of Calif. Leaflet* 2609. 5 pp. 12 color plates.

Nelson, H. D., W. W. Barnett, and C. A. Ferris. 1978. Fumigation of in-hull almonds on the farm. *Univ. of Calif. Coop. Ext.,* Fresno Co. Mimeo. 11 pp.

Rice, R. E. 1976. A comparison of monitoring techniques for the navel orangeworm. *J. Econ. Entomol.* **69**:25-28.

Rice, R. E. 1978. Unpublished report to Almond Board of California, Sacramento, CA.

Rice, R. E., and R. A. Jones. 1975. Peach twig borer: field use of a synthetic sex pheromone. *J. Econ. Entomol.* **68**:358-360.

Rice, R. E., L. L. Sadler, M. L. Hoffman, and R. A. Jones. 1976. Egg traps for the navel orangeworm, *Paramyelois transitella* (Walker). *Environ. Entomol.* **5**:697-700.

Roelofs, W., J. Kochansky, E. Anthon, R. Rice, and R. Carde. 1975. Sex pheromone of the peach twig borer moth. *Environ. Entomol.* **4**:580-582.

Soderstrom, E. L. 1977. Seal of almond shells and resistance to navel orangeworm. *J. Econ. Entomol.* **70**:467-468.

Summers, F. M. 1962. Insect and mite pests of almonds. *Calif. Agr. Exp. Sta. Circ.* 513. 16 pp.

Summers, F. M., and D. W. Price. 1959. Emergence and development of the overwintered generation of peach twig borer larvae. *J. Econ. Entomol.* **52**:340-341.

Thorp, R. W., and E. C. Mussen. 1977. Honey bees in almond pollination. *Div. Agr. Sci., Univ. of Calif. Leaflet* 2465. 3 pp.

Tumlinson, J. H., P. Sonnet, J. Coffelt, R. E. Doolittle, and K. W. Vick. 1978. Unpublished report to Almond Board of California, Sacramento, CA.

Wade, W. H. 1961. Biology of the navel orangeworm, *Paramyelois transitella* (Walker), on almonds and walnuts in northern California. *Hilgardia* **31(6)**:129-171.

Weir, W. C. 1951. Almond hulls as feed. *Calif. Agr.* **5(9)**:13.

Wood, M. N. 1947. Almond culture in California. *Calif. Agr. Ext. Serv. Circ.* 103 (Rev.). 87 pp.

WALNUT PEST MANAGEMENT: HISTORICAL PERSPECTIVE AND PRESENT STATUS

H. Riedl
Department of Entomology
University of California
Berkeley, California

M. M. Barnes
Department of Entomology
University of California
Riverside, California

C. S. Davis
California Cooperative Extension Service
University of California
Berkeley, California

INTRODUCTION

The Persian walnut, *Juglans regia* L., is the most important member of its genus as a commercial nut producer. Fifteen species of walnuts are native to North, Central and South America, the West Indies, China, Japan, and the Middle East. Several of these species are valued for their wood while others act as rootstocks for Persian walnut varieties. None match the Persian walnut as a producer of edible nuts.

Although the potential growing area of the Persian walnut includes other states where winters are mild, commercial plantings are found only in California, which produces more than 90% of the U.S. walnuts, Oregon, and southern Washington. Total walnut acreage has increased from *ca.* 140,000 acres in 1953 to over 200,000 acres in 1976. Walnut production has grown, even more dramatically, from *ca.* 60,000 inshell tons in the mid-fifties to 185,000 tons in 1976. This is related in part to new heavy-bearing varieties and denser plantings. With an upward trend in domestic walnut consumption and a significant expansion of the export market in recent years, a moderate but consistent growth is predicted for the walnut industry. The relatively stable price situation and the favorable economic prediction for this commodity undoubtedly will favor further growth in walnut acreage.

The walnut ecosystem provides habitat for a diverse complex of arthropods, nematodes, and pathogens. More than 25 insects and mites may cause injury to walnuts. Four feed directly on the nut or husk, whereas the majority feed on the vegetative parts of the tree above ground. There are six principal plant pathogens which attack either limbs, trunk, or the root system; one damages the developing nuts directly. Several nematodes infest walnut roots with the root-lesion nematode the principal species. This pest complex annually causes a significant reduction in yield and nut quality. Figures on yield loss are difficult to obtain, particularly for some chronic diseases and nematodes whose presence and effect are often not easily recognized.

As with other orchard crops, standards for quality and appearance are high. Kernels are graded by color, the lighter shades receiving a premium price. Walnuts sold in-shell as U.S. grade No. 1 are not allowed to have more than 5% internal damage from insects. External defects, such as shell staining may not be present on more than 5% of the nuts in a shipment.

Key pests such as the codling moth, the navel orangeworm, the walnut husk fly, and the bacterial pathogen causing walnut blight, may render a large portion of a crop unmarketable unless controlled by chemical means. However, the unwanted side effects of the single-factor chemical control approach were well recognized at an early stage on walnut with a research approach leading to lower dosages of DDT for codling moth control to minimize suppression of beneficials (Michelbacher and Bacon 1952a). In this paper we find an early use of the term "integrated control". These beginnings have led to the development of a successful pest management program where chemical control of codling moth is well integrated with biological control of the walnut aphid and successful chemical and cultural control of other pests (Barnes *et al.* 1978).

The authors purposely limit this review of walnut pest management to California since almost all the walnuts produced in the United States come from this state. It is perhaps for this reason, the lack of national importance of this commodity, that there has not been a review of this subject. In the light of recent advances and innovations in pest management on other crops *e.g.* alfalfa in Indiana (Giese *et al.* 1975) and apple in Michigan and other states (Croft 1975a) it is desirable to evaluate the status of pest management on walnuts critically and determine new directions and areas of emphasis.

This review focuses on the considerable diversity in terms of varieties, climate, soil and cultural practices under which walnuts are produced in California and how this affects pest management programs.

Pest management is not an isolated activity, rather it is an integral part of crop production. The authors emphasize, therefore, the tree and its management, and pests on walnuts should be studied in relation to crop management and various cultural activities. Each of these activities may modify directly or indirectly, in a positive or negative way, the life system of the many microbial and arthropod pests which live on walnut.

ORIGIN OF THE PERSIAN WALNUT AND HISTORY OF WALNUT CULTURE IN CALIFORNIA

The Persian walnut was known to the ancient Greeks. Its common name suggests Persia as its native home but its original distribution was larger and extended from the Caucausus to Turkestan (Chandler 1957). From Europe the Persian walnut reached North America with the early settlers during the 16th and 17th centuries. The term English walnut originated because walnuts were transported in English ships and became identified with England, though not an English export. This name also served to distinguish them from the native American walnuts. Today both common names are used, although Persian walnut is preferred.

There are several walnut species native to North America (Chandler 1957). The eastern black walnut, *Juglans nigra* L., is found east of the Rocky Mountains from southern Ontario to northern Florida. This tree, once plentiful in the eastern forests, has become very rare in the wild because of its sought-after furniture wood. Today it is grown primarily in plantations for veneer production (Anon. 1973b). The eastern black walnut is also a prolific producer of very flavorful nuts which are used in the candy industry. However, it is difficult to remove the kernels cleanly from the shells because of their hardness. The butternut, *Juglans cinerea* L., has a distribution similar to *J. nigra* but is more cold resistant and does not extend as far south. Its wood is less valuable than that of *J. nigra* and its nuts are even harder, though with a very flavorful nutmeat. Two *Juglans* species are native to California: the northern California black walnut, *Juglans hindsii* Jeps., and the southern California black walnut, *Juglans californica,* S. Wats.. The northern species is a large tree and serves as rootstock in many *regia* plantings.

The Persian walnut appeared in California with the early Spanish settlements, and they were frequently planted near Franciscan missions. These introductions came either directly from Europe or via Chile. The first large-scale plantings began around 1870 in southern California. Initially, walnuts were grown primarily from seedlings which came from parent trees with superior characteristics. Later, seedling trees were replaced by specific selections which were propagated on rootstocks leading to more uniform nut quality, size and yields. Varieties such as 'Placentia', 'Ehrhardt', and 'Chase' were early selections which were well adapted to the coastal climate in southern California (Rizzi 1977).

Walnut growing in northern and central California required varieties which were both frost-hardy and tolerant of high summer temperatures. Varieties imported from France were first established and better adapted to northern California, and of these introductions 'Franquette' became the most widely planted variety, but it has been reduced in acreage in recent years by newer more fruitful selections (Rizzi 1977).

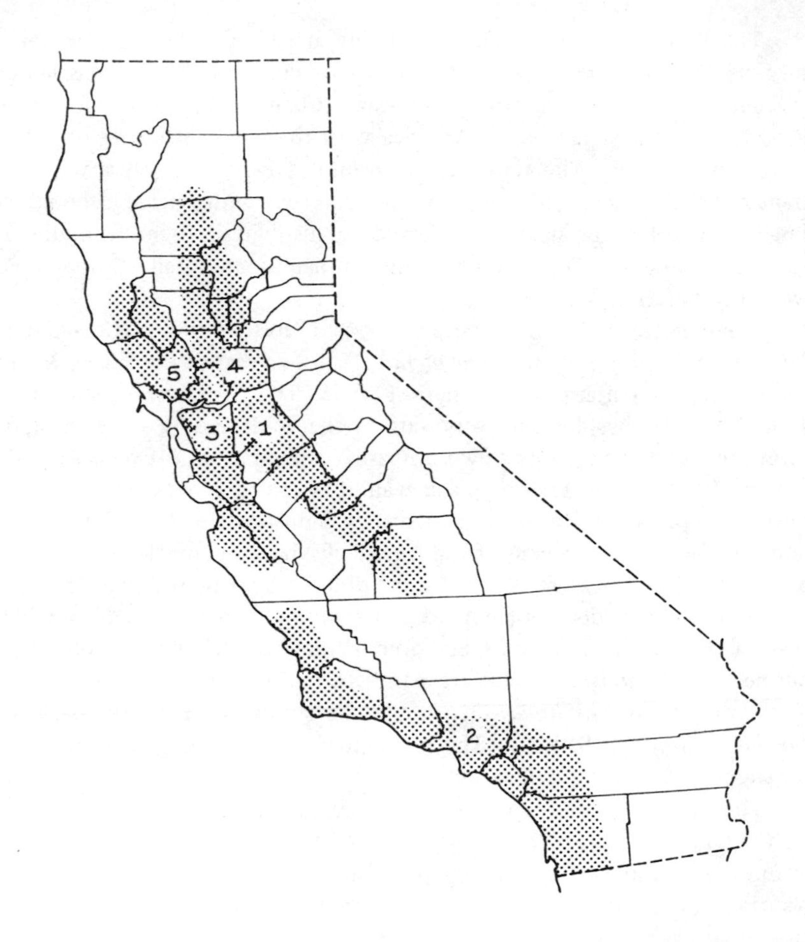

Figure 1. The major walnut-growing areas in California; Area 1: San Joaquin valley, 2: South coast, 3: Central coast, 4: Sacramento valley, 5: North coast.

Table 1.
Distribution of walnut acreage in California: 1953 versus 1976.

Area[a]	1953 Acreage[b]	1953 %	1976 Acreage[b]	1976 %
1. San Joaquin Valley	40,516	29	105,778	52
2. South coast	38,562	28	6,840	3
3. Central coast	28,551	21	14,782	7
4. Sacramento Valley	17,459	13	60,714	30
5. North coast	11,100	8	14,893	7
Other	1,426	1	2,602	1
State Total	137,614		205,609	

[a]See map, Fig. 1.

[b]Includes non-bearing and bearing acreage; based on data from Rock and Rizzi (1955) and the California Crop and Livestock Reporting Service.

In 1976 walnut acreage in California totaled 205,609 acres distributed over 43 of the 57 counties in this state. The main growing areas are indicated in Fig. 1. The centers of walnut production have markedly shifted over recent decades from the coastal counties in southern California around Los Angeles and counties east and south of the San Francisco Bay to the Sacramento and San Joaquin valleys (Table 1).

This trend is continuing and the expansion of walnut acreage in the two central valleys, areas 1 and 4 in Fig. 1, has not leveled off yet. The decline of walnut growing in areas 2 and 3 is primarily due to urbanization and the increase in land value. The expansion of walnut acreage in the two central valley areas more than makes up for the decline in other areas (Table 1).

THE WALNUT ECOSYSTEM

The Tree

Most *Juglans regia* varieties are large, spreading trees which are monoecious with staminate flowers (catkins) and pistillate flowers on the same tree. Pollination is by wind with little self- or cross-incompatibility among *regia* varieties.

Climate. The Persian walnut is not very cold hardy which limits its northern and altitudinal distribution. Optimal climatic conditions for nut production are rather specific. This is one reason why California produces the majority of walnuts in the United States since its climate is best suited for this crop. Pistillate flowers occur on the current season's shoots and are very susceptible to spring frosts. Fall frosts can destroy vigorously growing shoots, particularly on young trees. Cold resistance develops very slowly in the fall in most *regia* varieties which accounts for the susceptibility to early low temperatures. However, a certain amount of winter chilling, *ca.* 800 hours below 45°F, for some varieties is necessary for uniform bloom and good fruit set (Rizzi 1977). In southern California because of warm winters there is a frequent deficiency in chilling resulting in poor crops in late-leafing varieties. This is another reason the walnut acreage has declined in southern California and expanded in the north (Chandler 1957).

Summer heat can also be a problem. Nuts exposed to the sun may be damaged by radiant heat when air temperatures rise above 100°F, resulting in sunburned nuts. This may occur particularly in the interior growing areas of California where temperatures above 100°F in summer are quite frequent.

Few walnut regions in California have adequate rainfall for commercial production without irrigation (Table 2). The walnut tree requires between 28 and 38 inches of water during a full growing season. Depending on location and season, 1/2 to 3/4 of this amount has to be supplied by irrigation (Batchelor 1929; Fereres 1977). Most of the precipitation in this region occurs when trees are dormant. Late spring and summer rains can cause severe epidemics of walnut blight, while fall rains can interfere with harvest and cause mold problems in mature walnuts. The threat of walnut blight may be one reason walnuts are not grown commercially in the warm and humid regions of the southeastern United States.

Under the arid conditions of California, water is often the limiting factor for tree growth. Water stress can cause partial defoliation, shrivelled and off-colored nutmeats, and predispose the tree to attack by certain pests, particularly disease organisms. If a cover crop is maintained, the total water and nutrient requirement of the orchard increases, as well as frost hazard because of lower radiant heat reaching the tree from the soil. An orchard with a full cover crop needs about 1-1/2 to 2 times as much water as a weed-free orchard. During drought periods such as 1975-77, few growers maintained permanent cover crops.

In summary, the walnut does best in areas with little spring or fall frost, with winter rains and adequate winter chilling, with few days during summer above 100°F, and with a long growing season.

Soil. Few deciduous fruit trees in California are as sensitive to poor soil conditions, water salinity and high water table as the Persian walnut (Batchelor 1929; Begg 1977). Well-drained, deep, loamy soils with good aeration will

Table 2.
Normal precipitation and normal July temperature
in the five major walnut-growing regions of California.

Area[a]	Location	Precipitation[b] (inches)	Average July temperature (°F)[b]
1	Stockton	14.2	76.7
1	Fresno	10.2	80.6
1	Bakersfield	5.7	83.9
2	Santa Barbara	15.5	65.1
2	Los Angeles	14.1	72.0
2	Riverside	10.0	76.0
3	Hollister	13.3	66.5
3	San Jose	13.7	68.4
3	Livermore	14.5	71.3
4	Red Bluff	22.1	82.3
4	Marysville	21.0	78.8
4	Sacramento	17.2	75.2
5	Ukiah	38.4	72.9
5	Santa Rosa	30.5	66.8
5	St. Helena	35.4	70.8

[a]See map, Fig. 1.
[b]Averages from 1941-70; data source: NOAA.

produce vigorous trees and best yields. The Persian walnut grows poorly on shallow, sandy soils with a higher incidence of sunburning on the developing nuts. Clay soils with poor drainage are less suitable for walnut production and are, in addition, difficult to irrigate properly and are prone to develop crown rot. A permanent water table acts as a barrier to root development, but even a temporary, perched water table is damaging. Furthermore, the Persian walnut is sensitive to high concentrations of alkali salts. Therefore, the quality of the irrigation water is very important since it may cause or worsen a salinity problem of the soil. On the other hand, irrigation with salt-free water can help wash alkali salts from the root zone and thus improve growing conditions. Alkali injury shows up as browning of leaf margins.

Seasonal phenology. Once the leaves begin to appear in early spring, shoot and leaf growth is very rapid and most of the foliage has expanded to its ultimate size by the time pistillate bloom begins. Under optimal growing conditions a second flush of growth occurs during summer.

Most *regia* varieties are protandrous - that is, the male flowers begin to shed pollen prior to pistillate bloom. Some varieties are protogynous which makes them valuable pollinators. Varieties such as 'Placentia' and the newer 'Payne'-type selections have a shorter dormant period (November until March) while the old French varieties such as 'Franquette' are dormant from November until April. Varieties can differ significantly in their seasonal phenology as indicated by leafing date and harvest time (Table 3). This allows growers to choose varieties which best fit the specific climatic conditions of an orchard location.

Following pollination the nut goes through a very active growth period. Eight to ten weeks after pollination both nut and hull have completely sized (Fig. 2) and maturation of the kernel begins.

Rootstocks. At one time *J. californica* was tried as a rootstock but was soon discarded because of susceptibility to *Phytophthora* crown and root rot. Today *regia* varieties are propagated principally on two rootstocks: *J. hindsii,* the northern California black walnut and 'Paradox', a cross between *J. regia* and *J. hindsii.* Disease resistance of these rootstocks is discussed later in the section on rootstock susceptibility.

Varieties. California's walnut growing area is quite diverse in terms of climate (Table 2), soil type and topography. No variety could optimally fit all the combinations of growing conditions. Varieties which were well adapted to southern California (area 2) such as 'Placentia' did not perform well in the

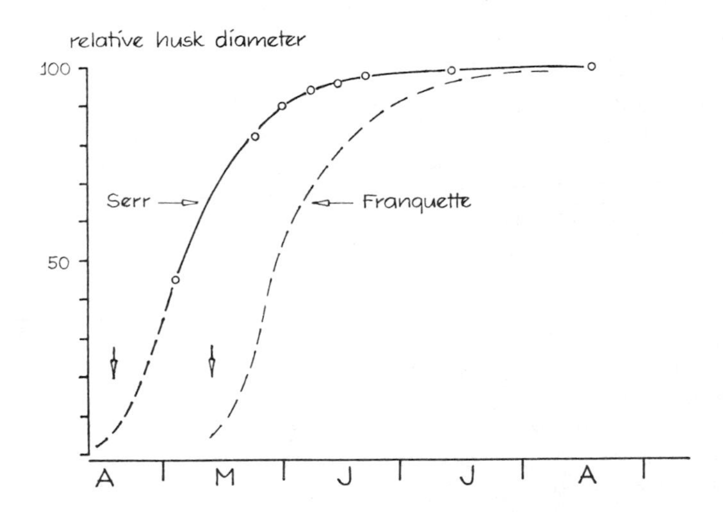

Figure 2. Seasonal pattern of nut growth in an early and a late variety (Based on data by D. E. Ramos, Coop. Ext. Ser., Univ. of Calif., Davis; arrows indicate peak bloom).

Table 3.
Varietal characteristics and susceptibility to several diseases and arthropod pests.[a]

Variety	Time of leafing (days after Payne)	% Lateral fruitfulness	Harvest period[b]	Shell seal[c]	Diseases[c]		Arthropod pests[d]	
					Walnut blight	Deep bark canker	Codling moth	Walnut husk fly
Ashley	0	90	E	G	S		S	R
Eureka	8	0	M	E	S		R	S
Franquette	30	0	L	G	R		R	S
Hartley	16	5	M-L	F	R	S	R	S
Payne	0	80	E	G	S		S	S
Serr	0	50	E	F-G	S		S	
Tehama	10	80	M	E-P	S		S	S
Chico	3	90	E-M	G	R		S	S
Vina	8	80	E	G	R		S	
Amigo	13	80	E	F	S		S	

[a]Modified from: Hendricks and Forde (1977a).
[b]E = Early, M = Midseason, L = Late
[c]P = Poor, F = Fair, G = Good, E = Excellent
[d]S = Susceptible, R = Resistant

interior valleys of northern California. The varieties which could be grown successfully in northern and central California were mostly the French introductions, 'Franquette' and 'Mayette', and their descendents. The late leafing and blooming habit of these varieties allows them to escape spring frost injury. They are generally light bearers and produce nuts primarily on terminal buds. More recent selections are more fruitful, producing nuts also on lateral buds. Table 3 compares several characteristics of old and new varieties such as time of leafing, fruitfulness of lateral buds, harvest period, etc. Phenological characteristics such as time of leafing and harvest period are indicators of frost susceptibility in spring and fall.

There has been a marked shift from the old light-bearing varieties to new 'Payne'-type heavy-bearing selections over the past two decades (Table 4). 'Franquette', at one time the most popular variety in the valley regions, is slowly being replaced by new introductions such as 'Ashley', 'Serr', and 'Tehama'. 'Placentia', once the leading southern variety, has almost disappeared because of loss of walnut acreage along the south coast (Table 1).

Table 4.
Major walnut varieties
in California: 1953 versus 1976.

Variety	% of total acreage[a] 1953	1976
Ashley	–	7.3
Concord	6.4	.5
Eureka	12.2	9.3
Franquette	27.2	16.1
Hartley	4.7	28.6
Mayette	4.2	1.6
Payne	17.2	16.5
Placentia	18.1	.4
Serr	–	7.6
Tehama	–	2.9
Others	10.0	9.2

[a]Based on data from the California Crop and Livestock Reporting Service.

Varietal selections have been made on the basis of leafing date, their adaptability to specific climatic conditions, lateral bearing habit, vegetative vigor, harvest timing, and quality traits concerning the kernel and shell (Serr and Forde 1968; Hendricks and Forde 1977a, b) but little attention has been paid to disease and arthropod pest resistance in these breeding programs.

Planting. In most older orchards tree densities ranged from 12 to 27 per acre. A spacing of 60 feet was considered best to accommodate the large trees of the light-bearing older varieties. These large trees were difficult to manage. Many of today's varieties are smaller trees and can be planted closer to increase production. Spray coverage on large walnut trees has always been a problem, even with specialized equipment. In the newer high-density plantings spray operations are more efficient and easier to carry out because of smaller tree size. Therefore, for an efficient pest control operation throughout the life of the orchard, eventual tree size, which is determined by variety and spacing, needs to be considered.

Soil management. Some form of soil management is necessary to provide for optimum tree growth and to create an orchard floor accommodating spraying, irrigation and the mechanical harvesting operation. Choice of a particular management system will be determined by factors such as soil type, irrigation system, water availability, water and fertilizer costs, orchard climate and pest problems (Osgood 1977; Elmore and Lange 1977).

When complete cultivation is practiced the orchard is disked three to five times a year (Fig. 3A). This method is well adapted to various irrigation methods and creates a warmer orchard climate by virtue of heat radiation from the soil. The latter may be important in areas with spring and fall frost hazards. Disadvantages are compaction on certain soil types, poor water penetration and dust problems which may induce mite build-up. Root and lower trunk injuries may be caused by disking and deep tillage and result in crown gall infections.

With complete sod culture, soil compaction does not occur as readily since plant roots maintain good water movement. This system is cooler than complete cultivation which may be an advantage during hot summer days but a disadvantage during days of frost danger. Various cover crop systems adaptable to walnuts are described by Finch and Sharp (1976). Fewer mite problems are associated with sod since the ground cover diminishes dust problems. However, if not properly managed, particularly when a volunteer cover of certain plants develops, mites may build on morning glory for example, and spread into the tree. This could be avoided by development of information on cover crops which are not conducive to pest problems. The biggest drawback of cover cropping is the additional expense for water and nitrogen. In years of water shortage, it is imperative to switch to less water-intensive management systems.

Figure 3. (A) Freshly-disked walnut orchard being readied for irrigation; (B) Blight lesion on walnut caused by *Xanthomonas juglandis.*

Complete chemical weed control is adapted well to sprinkler irrigation, avoids soil compaction, but may cause heat problems in the summer. On heavy soils there may also be problems with water penetration. A residual herbicide is applied followed by spot treatment with a contact herbicide.

Sod culture and strip chemical weed control in the tree rows combines several advantages. If the cover crop is kept short, an orchard with this system will be only slightly cooler than a completely cultivated orchard thus reducing frost hazard. From a pest management standpoint this soil management system may have advantages. Cultivation equipment-related injuries to the crown area leading to crown gall infections do not occur as readily. In addition, the sod cover between the tree rows increases diversity in the orchard which may also increase the stability of pest-natural enemy interactions in the tree.

Irrigation. Depending on the growing area, walnut trees require between 28 and 38 inches of water for normal production (Fereres 1977). Water need is highest in the interior valleys (area 1 and 4, Fig. 1) where humidity is lower and temperatures higher than in the coastal growing districts (areas 2, 3 and 5). Normal rainfall is adequate only in exceptional locations along the north coast (Table 2), but even there because of lack of rain during late spring and summer, irrigation may be necessary. It is important to monitor soil moisture throughout the year, even during the dormant period. Winter irrigation may be necessary in districts 1 and 4 to make sure the tree is fully supplied with water throughout the root zone when growth begins in spring. Inadequate availability of water during spring and early summer, the main growth period, will result in smaller nuts, lower yields and poor quality. Mid-summer or late irrigations have little effect on eventual nut size but do affect kernel quality. If the soil is too dry after harvest, young trees become susceptible to fall frost. Post-harvest irrigations will remedy this situation but may lead to crown rot infections if trunks are exposed to standing water for longer time periods (Meyer *et al.* 1975).

Harvesting. Nuts are normally harvested at 80% husk split (first shake); the remaining nuts are harvested about 10 days later (second shake). If harvest is delayed, or if nuts remain on the ground too long, kernel quality (primarily color) goes down due to sun exposure and high temperatures, and damage from the navel orangeworm and molds increases (Sibbett *et al.* 1972 and 1974). The mold problems in mature nuts are aggravated if fall rains have begun.

Kernel and husk usually mature at the same time in the walnut districts along the coast (areas 2, 3, and 5), while in the interior valleys (areas 1 and 4) the kernel may mature two to three weeks ahead of the husk. Once the kernel is mature, quality deterioration begins, its rate being governed by temperature. To avoid continued sun exposure which darkens the meats and lessens quality, hull split can be promoted by applications of ethephon, a growth regulator discussed later in the chemical control section. This practice, together with prompt harvest, dehydration, and fumigation not only increases nut quality but also obviates the need for cultural control or orchard use of chemical control of navel orangeworm (Sibbett *et al.* 1974).

Pest Complex on Persian Walnuts

The walnut in California is host to several pathogens infecting various plant parts, root-inhabiting nematodes and many arthropods which may feed on leaves, nuts or limbs (Table 5). Diseases and nematodes will not be treated in great detail and the reader is referred to recent publications in these subject areas. Weeds are discussed in the section on chemical control.

Diseases. Several pathogenic fungi and bacteria may cause severe disease problems in commercial orchards resulting in reduced yields and damage to the tree.

Fungal pathogens affecting the root system, the crown, and the lower trunk are *Armillaria mellea* commonly called oak root fungus, and *Phytophtora* sp. causing root and crown rot. Both are soil-borne fungi which can persist in the soil for many years.

Fruit and nut trees as well as many other kinds of deciduous trees are attacked by the fungus *(Armillaria).* First the roots are infected, then the crown region, and eventually the lower trunk is completely girdled often resulting in tree death. Symptoms are stunted terminal growth, leaf drop and dieback of terminals. The disease can be recognized by fan-shaped yellowish patches of mycelium below dead bark at the soil line and black strands of rhizomorphs on root surfaces. *Armillaria* infected roots are spread by equipment to clean areas in an orchard where they come in contact with healthy roots. Resistant rootstocks are the principal means of controlling this disease, as discussed later (La Rue *et al.* 1962; Mircetich *et al.* 1977a).

Several species of the fungus group known as *Phytophthora* root and crown rots are associated with walnuts in California. Infection will result in reduced terminal growth, sparse foliage, yellowing of leaves, girdling of the trunk and eventual collapse of the tree. Moist, cool conditions in the spring and fall are very conducive to infection and promote disease progression. The site of infection is often the crown near the soil line. Keeping the crown area dry is one way to prevent infection. Spread of the disease occurs primarily during surface irrigation. Resistant rootstocks and good water management are the principal and the most economical ways to control root and crown rot (Rackham and O'Reilly 1967; Mircetich *et al.* (1977a).

The causal agent of crown gall is *Agrobacterium tumefaciens,* a soil-borne bacterium which forms galls on the crown or the roots below the soil

Table 5
Diseases, nematodes and arthropods affecting various parts of the walnut tree.

	Indirect Pests			Direct Pests	
	Root system	Trunk & limbs	Leaves	Husk	Kernel
Diseases	Armillaria Crown rot Crown gall	Deep bark canker Shallow bark canker Blackline	Walnut blight	Walnut blight	Molds
Nematodes	Root-lesion Root-knot Other species				
Arthropods		*Coleoptera* Woodborers *Homoptera* *Armored scales:* Oystershell Italian Pear Putnam Walnut San Jose *Soft scales:* Frosted Fruit Lecanium Calico *Hemiptera* False chinch bug	*Lepidoptera* Redhumped caterpillar Fruittree leaf-roller Fall webworm Walnut spanworm *Homoptera* *Soft scales:* Frosted Fruit Lecanium Calico *Aphids* Walnut Dusky-veined *Acarina* Twospotted mite Pacific mite European red mite	*Diptera* Walnut husk fly	In Orchard: *Lepidoptera* Codling moth Navel orangeworm Filbert worm In Storage: *Lepidoptera* Indian meal moth Mediterranean flour moth *Coleoptera* Sawtoothed grain beetle

line. The bacterium enters the plant only through wounds. Infection frequently occurs in the nursery. In established orchards the bacterium invades the tree through wounds caused by cultivation equipment. Trees girdled by crown gall suffer from stunted growth (Ross *et al.* 1970).

The nature of the disorder called blackline appears to be a graft-transmissible infectious agent. A narrow line of dead tissue forms in the graft union between rootstock and scion which eventually leads to complete girdling and death of the scion. Disease symptoms are similar to the other root diseases. Blackline has been more frequent in the central coast growing district (Area 3) but is now spreading in the Sacramento and San Joaquin valley. 'Paradox' and *J. hindsii* rootstocks are affected by this disorder while *J. regia* rootstocks do not display this graft problem (Serr 1959; Martin and Forde 1975; Mircetich *et al.* 1977b).

Deep bark canker or phloem canker is caused by the bacterium *Erwinia rubrifaciens.* Bark cracking and oozing of lesions on trunk and scaffold branches are typical symptoms in late summer. Dissemination is achieved by wind-blown rain and also by shaking equipment which can transmit bacterial ooze from tree to tree during the harvest operation. This disease is often associated with poor growing conditions. Infected trees are able to recover if they regain vigor. The 'Hartley' variety is most susceptible, particularly in the warm valley districts (Schaad and Wilson 1971; Schroth *et al.* 1977).

Another bacterial disease of walnuts is shallow bark canker caused by *Erwinia nigrafluens.* Its effect on the trunk is superficial since lesions do not penetrate to the cambium. In most cases this disease is not considered very damaging (Schroth *et al.* 1977).

Walnut blight is by far the most widespread and troublesome. Infections occur only under moist conditions, and the hazard is greatest during the spring rains. Leaves, nutlets and tender shoots may develop dark lesions caused by the bacterial agent, *Xanthomonas juglandis* (Fig. 3B). This bacterium overwinters in infected buds, twig lesions and catkins. Dissemination of contaminated pollen is one way of pathogen spread. Rain and insects also vector this pathogen. Nuts infected early in the season drop while later infections may cause shrivelled kernels. Blighted nuts are subject to infestation by the navel orangeworm, leading to population build-up of this pest (Ark and Scott 1951; Olson *et al.* 1976; Schroth *et al.* 1977).

Nematodes. Many walnut trees in California have nematodes associated with them. These organisms parasitize the feeder root system and weaken the tree and reduce yield and quality. The root-lesion nematode, *Pratylenchus vulnus,* is the most serious nematode pest in walnut orchards (Lownsbery and Sher 1958). It produces black lesions in the roots and destroys the feeder roots. Another nematode which is found in walnut orchards and which may attack the feeder root system is the ring nematode, *Criconemoides xenoplax.* Problems with nematodes are common in replant orchards but are often not recognized. 'Paradox' rootstocks show greater tolerance to the root-lesion nematode (Lownsbery 1977).

Arthropod pests. The only list of arthropod species associated with the genus *Juglans* has been provided by Barrett (1932) based on records in the literature. Of the 336 species in this survey, 120 reportedly occurred on Persian walnuts in California. The species complex which inhabits the Persian walnut in the state is probably richer than indicated. Nixon and McPherson (1977) found approximately 300 phytophagous insects in ten orders on immature black walnut, *J. nigra* in southern Illinois. Since Barrett's survey, ten additional arthropod species have been described on walnuts, some of which have become serious pest problems (Michelbacher and Ortega 1958). However, Barrett's (1932) prediction that several serious pests of native walnuts in the eastern United States would soon arrive in the walnut-growing areas of California has fortunately not materialized.

Today over 25 arthropods are of economic significance on walnuts in California. More than one-third of those are introductions into the state including the codling moth, navel orangeworm, several scale insects, aphids and the walnut husk fly.

The following is a brief description of major arthropod pests, their biologies and their interactions with the host tree. For a more detailed account the reader is referred to Michelbacher and Ortega (1958). The walnut pest complex can be categorized into direct and indirect pests according to feeding habits and the plant parts attacked (Table 5). Within these two broad categories pests can be grouped according to their economic status and severity and frequency of attack, into key, occasional and induced pests (Table 6). The number of pests in each of these groups and their damage potential differ regionally due to climatic and growing conditions, variety, and cultural practices.

Arthropod pests (Key Pests). The codling moth, *Laspeyresia pomonella* (L.), a worldwide pest of pome fruits, is the most serious insect pest of early season, heavy-bearing varieties of walnuts. In northern California, the codling moth was first reported on walnut in 1909, with significant infestations occurring only near heavily infested pear orchards and pear packing sheds, with light infestations elsewhere in Contra Costa County (Foster 1912). Light infestations in the northern counties, however, were also related to the use of French varieties, rather than failure to adapt to walnut.

Though apples and pears had been heavily infested in Riverside County in southern California since the 1880's, only light infestations of negligible

Table 6.
Timing of control measures
for various walnut pests in relation to tree phenology.[a]

	Seasonal phenology of walnut tree (Payne variety; Davis, CA)[b]	Jan-Feb	Mar	Apr	May	Jun	Jul	Aug	Sep	Oct	Nov	Dec
		dormant	leafing-out	peak bloom					hull-split		leaf-drop / dormant	
Key Pests	Walnut blight		X	X	X							
	Codling moth			X[c]			X					
	Navel orangeworm											
	Walnut husk fly						X[c]	X				
Occasional Pests	Dusky-veined aphid			X			X					
	Filbert worm											
	Redhumped caterpillar						X	X				
	Fruittree leafroller											
	Fall webworm											
	Walnut spanworm											
Induced Pests	Oystershell scale	X				X[c]						
	Italian pear scale	X				X[c]						
	Putnam scale	X				X[c]						
	Walnut scale	X				X[c]						
	San Jose scale	X				X[c]						
	Frosted scale	X	X[c]			X[c]						
	Fruit lecanium	X	X[c]			X[c]						
	Calico scale	X	X[c]			X[c]						
	Walnut aphid			X			X					
	European red mite	X					X	X[c]				
	Pacific mite						X	X[c]				
	Twospotted mite						X	X[c]				

[a]Based on Univ. of Calif. pest control recommendations for walnuts (Barnes et al. 1973).

[b]Based on records from D. E. Ramos, Agric. Ext. Service, Univ. Calif., Davis.

[c]Optional to other timing.

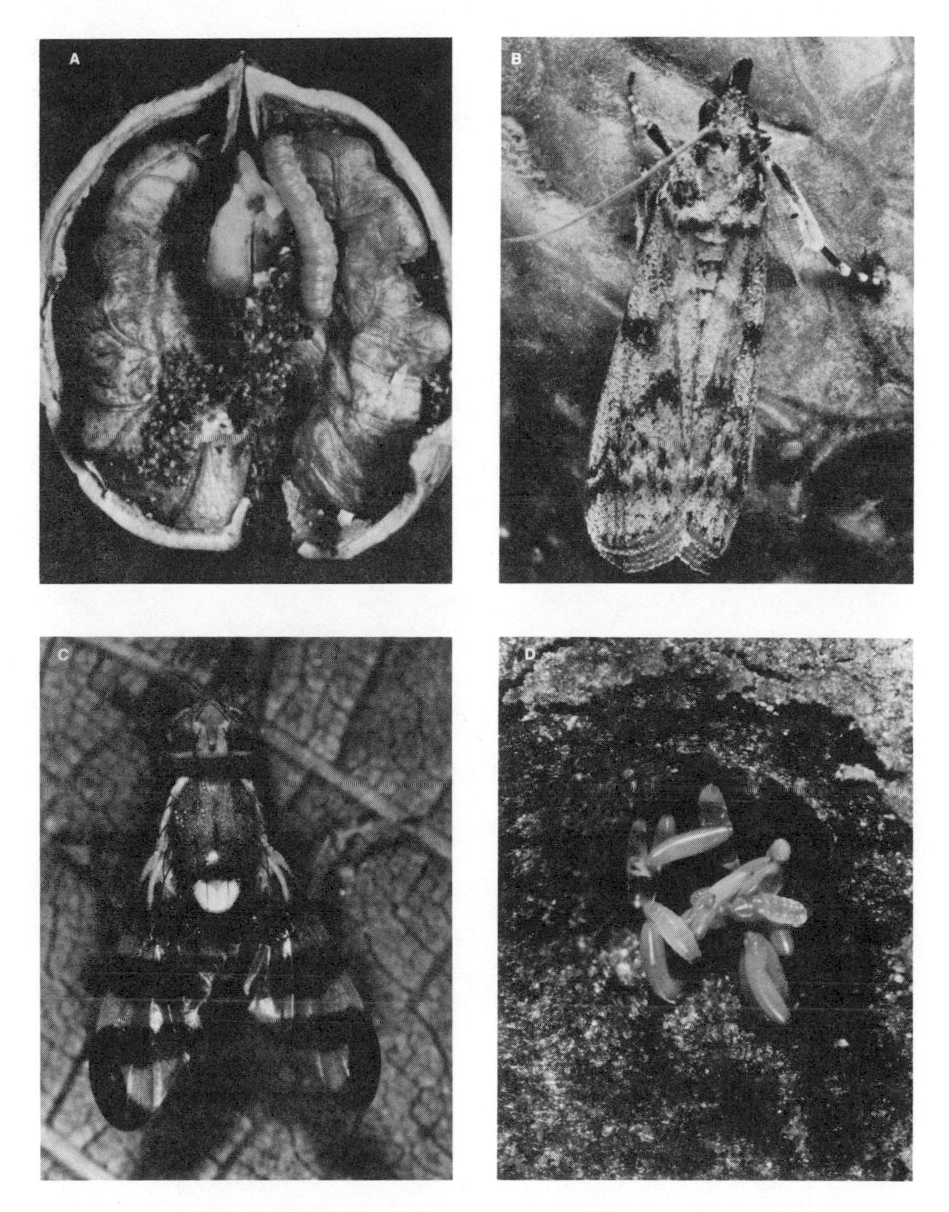

Figure 4. (A) Codling moth larva feeding on the kernal; (B) Navel orangeworm adult (Courtesy of L. Dunning, Univ. of Calif., Davis); (C) Walnut husk fly adult (Courtesy of J. Clark, Univ. of Calif., Davis); (D) Walnut husk fly eggs and freshly hatched larvae (Courtesy of J. Clark, Univ. of Calif., Davis).

importance on susceptible varieties of walnut were reported until 1931 (Boyce 1935). This slow development of codling moth as a walnut pest was also reported for coastal counties in southern California by Quayle (1926). The codling moth is now well adapted to walnuts in all areas of the state and has apparently formed a distinct host race. Moths of the walnut race prefer to lay eggs on walnut foliage as compared with apple foliage (each with associated fruit) and this trait has a genetic component (Phillips and Barnes 1975). The preferred site for larval entry in early season is the calyx. Later, larvae enter through the side or base of a nut. Before the shell hardens, larvae tunnel directly to the center and feed on the kernel (Fig. 4A). Later, the only possible entry point is through the fibrous tissue of the suture at the stem end of the nut.

There are at least two full and often a partial third generation per year (Barnes 1978; see also Fig. 5). The presence and size of the third generation depends on latitude, coastal influence and the climatic conditions of a given season. On susceptible varieties, damage levels at the end of the first generation can reach 30%, and 50% or more after the second generation (Michelbacher and Ortega 1958).

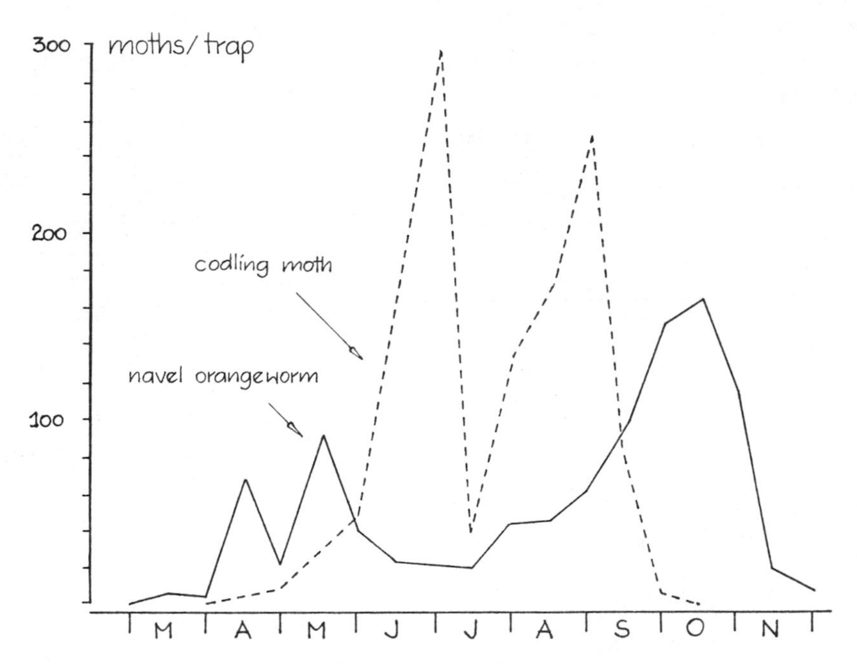

Figure 5. Seasonal light trap catches of the codling moth and the navel orangeworm at Walnut Creek, Calif. (Data from Madsen and Wong 1962).

As in other deciduous orchard regions where the codling moth occurs, natural enemies are unable to regulate populations below tolerable levels. Under moist conditions high mortality of overwintering larvae from a *Beauveria* fungus has been observed (Michelbacher *et al.* 1951). Egg parasites, *Trichogramma* species, can frequently be found in the field but their impact is negligible. *Ascogaster carpocapsae* Viereck, a small braconid wasp, attacks the egg but emerges after the larval damage is done. It is fairly common but rarely reaches high parasitization levels. Recently an ichneumonid parasite, *Liotryphon caudatus* (Provancher), has been introduced from Afghanistan and released on a limited scale (Etzel 1978). No active foreign exploration for natural enemies is underway at this time.

The navel orangeworm, *Amyelois transitella* (Walker), was first described in cull oranges in Mexico. In 1943 it was collected from walnuts in southern California. This pest is a scavenger and has spread to the major nut-producing areas of the state. The larvae scavenge a wide range of fruits, nuts and legume pods, almost all of which are dried and adhering to the bearing plant (*i.e.* mummified), previously infested by other insects, decayed, or otherwise damaged. Over 40 scavenged host materials have been recorded. However, the navel orangeworm infests and develops well on sound new crop walnuts, almonds and pistachios after the hulls split.

In spring, the silvergray moths (Fig. 4B) of the overwintered brood lay eggs singly on mummy walnuts left in the orchard, codling moth-infested, or blighted nuts. Sunburned nuts apparently do not invite infestation. Several larvae may feed at the same time in one nut and infestation is accompanied by heavy webbing. Pupation usually takes place inside the nut.

This insect has no diapause, it is multivoltine and overwinters in unharvested, mummy nuts on the tree and trash nuts on the ground. Provided codling moth and blight have been chemically controlled, the lowest population levels are recorded in early summer when the supply of nuts from the previous year is exhausted (Fig. 5). Populations increase following codling moth infestations, blight infections and hullsplit. Highest moth populations occur in early fall and represent an explosive increase following hullsplit and availability of trash fruits in adjacent orchards (Fig. 5). If a walnut orchard is not properly managed and harvest unduly delayed, this pest can become very destructive. Natural enemies are of little avail in suppressing populations of this insect below tolerable levels, although a few natural enemies have been reported (Michelbacher and Davis 1961a, b). Among those were several hymenopterous parasites including *Microbracon hebetor* (Say) and predators. A parasite has been introduced from Mexico, the polyembryonic encyrtid, *Pentalitomastix plethoricus* Calt.

(Caltagirone 1966). Large numbers of this parasite have been released in almond orchards but no detectable effect upon levels of infestation has occurred.

The walnut husk fly, *Rhagoletis completa* Cresson (Fig. 4C), is one of four indigenous *Rhagoletis* species which feed on native walnuts east of California. None of them occurred in California until *R. completa* was discovered in San Bernardino County in 1926 (Boyce 1934). From this initial infestation the husk fly spread throughout southern California and by 1954 reached the north coast growing district (Michelbacher and Sisson 1956). Today, the husk fly is an annual pest problem in all California walnut areas, except for the southern San Joaquin Valley, where it occurs primarily on back-yard trees. All varieties of Persian walnut as well as the native black walnut are suitable hosts.

In contrast to most other *Rhagoletis* species the husk fly lays its eggs in groups just below the surface of the husk (Fig. 4D). Larvae feed inside the husk tissue, and when mature drop to the ground to pupate and overwinter. The husk fly is univoltine with a one to four year diapause. On very susceptible varieties infestations can reach 90-100% at the end of the season. Larval feeding can cause various conditions which reduce quality such as discolored or shrivelled kernels, stained shells, nut molds, and prevention of clean separation of the nut from the hull. In recent years, the majority of the crop has been sold as nutmeats and shell staining by husk fly larvae in late season has ceased to be of such importance.

Boyce (1934) noted that the husk fly was remarkably free of natural enemies. In its native habitat, several hymenopterous parasites occur which attack *R. completa* and related species (Buckingham 1975). Presently an exploration and release program for husk fly parasites is underway in order to establish them in California's walnut districts (Hagen 1978).

Arthropod Pests (Occasional Pests). A number of walnut pests occur only sporadically and are of more regional importance. Because of their sporadic nature they can often surprise an unsuspecting grower and seriously damage the crop. Most of these pests are polyphagous and in some years become pests on walnuts in areas where their primary hosts occur.

In 1928 the dusky-veined aphid, *Callaphis juglandis* (Goeze), was first discovered in the Willamette Valley in Oregon. By 1952 it was reported from San Jose and soon thereafter its presence was confirmed from throughout northern California. In contrast to the walnut aphid, the dusky-veined aphid feeds on the upper side of leaves. Nymphs line up in a very characteristic manner along the midvein during feeding (Fig. 6A).

In recent years populations of *C. juglandis* have been on the increase. Lack of competition with the walnut aphid and the virtual elimination of

Figure 6. (A) Dusky-veined aphid nymphs feeding along the midvein and two winged female adults (Courtesy of W. H. Olson, Univ. of Calif., Coop. Ext. Ser.);
(B) Nuts and leaves covered with large amounts of honeydew due to walnut aphid feeding;
(C) *Trioxys pallidus* parasitizing a walnut aphid nymph (Courtesy of J. Clark, Univ. of Calif., Davis);
(D) Walnut scale: adult female with cover removed, note lateral lobes (Courtesy of J. Clark, Univ. of Calif., Davis).

any control measures against the same are believed to be responsible for the upward trend of *C. juglandis.* The damage to the crop resulting from feeding by this aphid is similar to that by the walnut aphid. Infestations during spring have the greatest effect on the crop but summer populations can be damaging as well (Olson 1974).

A complex of natural enemies, primarily predators, attack the dusky-veined aphid. Several new parasites are being introduced to improve biological control of this aphid (van den Bosch 1978).

The filbertworm, *Melissopus latiferreanus* (Walsingham), was of great concern in the mid 1950's in the Sacramento valley and along the north coast (Michelbacher *et al.* 1956 and 1957). However, there have not been any recent infestations. Among its hosts in California are acorns and especially the oak apple galls caused by *Andricus californicus* Barrett, a small gall fly. Larvae are able to enter the nut only after hull split. Should it re-occur, this pest can probably be controlled by early harvest practice.

Several lepidopterous defoliators can become walnut pests. Among these the redhumped caterpillar, *Schizura concinna* (J. E. Smith), appears to be the most destructive, particularly in the warm central valleys. If unchecked, the gregarious larvae can easily strip a small tree of all its leaves. Other fruit trees and ornamentals also serve as hosts. The fruittree leafroller, *Archips argyrospilus* (Walker), although a serious pest of many other orchard crops and shade trees, is of relatively minor importance on walnuts. So is the fall webworm, *Hyphantria cunea* (Drury), which rarely causes much damage. Spotty infestations occur at times throughout the interior valleys. The walnut spanworm, *Coniodes plumogeraria* (Hulst), normally feeds on live oak. Occasionally it attacks walnuts and other deciduous fruit trees. This pest is present throughout the Pacific Coast states, but outbreaks on walnuts have only been reported from southern California. In orchards with regular spray programs against the codling moth these defoliators rarely develop populations which require treatment.

The false chinch bug, *Nysius raphanus* Howard, is primarily associated with the volunteer cover crop and grassland, especially with London rocket, *Sisymbrium irio* L.. This pest moves into trees when its weed hosts dry up. Feeding results in partial defoliation and may kill tender twigs on young trees.

Besides walnuts, several species of Coleoptera in the families Buprestidae, Cerambycidae and Scolytidae, attack many forest, shade and fruit trees which are under stress, injured or diseased. Healthy trees with good vigor are rarely attacked. The Pacific flatheaded borer, *Chrysobothris mali* Horn, is perhaps the most destructive of this group. This borer is a cambium feeder and can girdle limbs or even trunks of small trees. Young orchards are especially susceptible.

Arthropod Pests (Induced Pests). Pests which are normally well regulated by their biological antagonists but which can be induced to pest status when the natural enemy complex is disturbed fall into this group. These include the walnut aphid, several soft and armored scales and phytophagous mites. Most often the spray programs applied against the key pests are responsible for secondary pest outbreaks because of their non-selective properties and high toxicity to the natural enemy complex. However, in some years when environmental conditions favor the pest over the natural enemies, pests may build up to damaging levels and corrective action is justified. If properly managed, these fluctuations will revert back to an equilibrium condition below the economic threshold.

From the outset of the development of the walnut industry in California, the walnut aphid, *Chromaphis juglandicola* Kaltenbach, was one of the most important pests. It was introduced from Europe and in the absence of effective natural enemies annually caused severe losses unless chemically controlled. This species is active from the time leaves appear until leaf drop and feeds on the underside of leaves. Populations are strongly suppressed by temperatures over 100°F which continue for several days. There are 8 to 10 generations per year. Populations build up quickly during spring when the nuts go through a very active growth period. Aphid feeding can cause premature leaf drop, reduce nut size and yield, and produce large amounts of honeydew (Fig. 6B). The honeydew kills the epidermal cells of the nuts, blackening them and leading to severe sunburn damage (Sibbett *et al.* 1971). Only the early varieties are susceptible to this type of damage, once mistaken for direct sunburn injury. It may be virtually avoided if aphids are under satisfactory management. Poor biological control by indigenous parasites and predators and problems of pesticide resistance prompted a search for new more effective parasites. In 1959 a French strain of *Trioxys pallidus* Haliday, a small braconid wasp (Fig. 6C), was introduced which soon controlled the walnut aphid in the cooler coastal growing areas but failed in the arid central valleys (Schlinger *et al.* 1960; van den Bosch *et al.* 1962). With the establishment in 1968-70 of a *T. pallidus* strain from Iran which was well-adapted to the warm interior of California, routine chemical control of the walnut aphid became unnecessary.

Several deciduous fruit trees are attacked by the frosted scale, *Lecanium pruinosum* Coquillett, but walnut appears to be one of the preferred hosts. In the spring the overwintering females produce a large number of eggs. Upon hatching the crawlers move to the young growth, leaves and twigs, where they feed for the whole season. Outbreaks of frosted scale in the late 1940's and early 1950's were related to the use of DDT for codling moth control which adversely affected the natural enemies of this scale (Bartlett and Ortega 1952; Michelbacher 1955). If undisturbed by pesticides several parasites, in particular *Metaphycus californicus* (Howard), a small braconid, provide effective control except for occasional failures.

Two other soft scales, the European fruit lecanium, *Lecanium corni* Bouche, and the calico scale, *Lecanium cerasorum* Cockerell, are similar to the frosted scale in biology and the kind of injury they cause, but are infrequently encountered. Both are effectively held in check by their natural enemies.

The armored scales which infest walnuts are primarily bark feeders. Their host range includes other deciduous fruit and also shade trees. The most common and important armored scale on walnut is the walnut scale, *Quadraspidiotus juglans-regiae* (Comstock) (Fig. 6D). Occasionally, in localized areas of the orchard, San Jose scale, *Quadraspidiotus perniciosus* (Comstock) is a problem. The oystershell scale, *Lepidosaphes ulmi* (L.), the Italian pear scale, *Epidiaspis piricola* (DeGuer.), and the Putnam scale, *Diaspidiotus ancylus* (Putnam), are less frequently encountered.

When infestations of these scales are heavy, tree vigor may be severely reduced and sometimes whole limbs may be killed. Normally these scales appear to be suppressed by natural control factors but outbreaks do occur. In recent years, the walnut scale has become a rather wide-spread pest in the central valley.

Phytophagous mites are pests with wide host ranges and similar feeding habits. Mite outbreaks were often associated with heavy use of DDT, repeated use of nicotine dusts and other organic insecticides (Michelbacher and Bacon 1952a, b). Extensive mite feeding can cause leaf browning, partial defoliation and reduce yields significantly (Barnes and Moffitt 1978). Mite populations usually increase during summer and early fall. Summer infestations of European red mite were shown to have no effect on yield until the third season. Also, after two years of infestation there was a severe reduction in staminate flowers. If a tree is defoliated as a result of heavy mite feeding, branches as well as the nuts are exposed to sun radiation and can suffer sunburn injury. A diverse complex of predators regulates the various phytophagous mites common to walnuts and can keep them at subeconomic levels. The most important beneficials are predaceous mites, the sixspotted thrips, *Scolothrips sexmaculatus* Pergande, the ladybird beetle, *Stethorus picipes* Casey, anthocorid bugs, and green lacewings.

Several plant-feeding mites are injurious to walnuts such as the European red mite, *Panonychus ulmi* (Koch), the twospotted spider mite, *Tetranychus urticae* Koch, and the Pacific mite, *Tetranychus pacificus* McGregor. Other less important mite pests are the carmine spider mite, *Tetranychus cinnabarinus* (Boisduval), the walnut blister mite, *Aceria erinea* (Nalepa), and the false spider mite, *Brevipalpus lewisi* McGregor. The citrus red mite, *Panonychus citri* (McGregor), may "balloon" from citrus orchards and establish populations on walnut.

Certain orchard conditions tend to invite and magnify mite problems. It has been observed that dust from roadsides and frequently from cultivation practice in orchards create mite outbreaks. A weedy orchard floor with morning glory, pigweed and other plants preferred by mites may lead to migration and infestation in the tree. Also, trees which are not properly irrigated tend to be more susceptible to mite damage.

THE PEST COMPLEX OF WALNUTS IN RELATION TO ROOTSTOCKS, VARIETIES AND GROWING REGIONS

Rootstock Susceptibility

J. hindsii rootstock provides for resistance to *Armillaria*, the oak root fungus, but is susceptible to crown rot, blackline, and the root-lesion nematode, *P. vulnus*. 'Paradox' is a vigorous rootstock which is variably resistant to *Armillaria*, resistant to root and crown rot, and somewhat tolerant of *P. vulnus*, but also susceptible to blackline. *J. regia* rootstock is very susceptible to *Armillaria* and sensitive to high salinity. However, because of its resistance to blackline, an expanding problem in many commercial orchards, *J. regia* may become a valuable alternative provided soil salinity is low and the oak root fungus absent (Mircetich *et al.* 1977a, b; Lownsbery 1977; Martin and Forde 1975).

Choice of a proper rootstock is one of the first pest management decisions a grower must make before the orchard is actually planted. Alternatives should be carefully considered based on an analysis of known rootstock problems in an area. Unwise choice may prevent the orchard from reaching its full production potential and even may cause early tree loss and make it impossible for a grower to recover his investment.

Varietal Susceptibility

The principal concern in walnut breeding programs has been to select varieties with climatic adaptability, high production potential and specific quality traits while little attention has been paid to insect or disease resistance. Several of these characteristics such as leafing date, time of pistillate and staminate bloom, bearing habit, harvest period and shell seal are variously related to resistance to certain walnut pests (Table 3).

In spite of its importance few experimental studies have dealt with varietal resistance to specific pests and consequently there is a paucity of information on this subject.

One of the principal and very obvious resistance mechanisms in walnuts is phenological asynchrony between a variety and a pest. This mechanism is also referred to as ecological resistance since it is primarily controlled by environmental factors. Other genetically controlled resistance mechanisms such as phenetic resistance, antibiosis and tolerance seem to be important on walnuts but they are not well documented.

The following is a brief summary of varietal resistance based on published experimental work, field observations and inferences drawn by the authors from related studies.

Deep bark canker is particularly serious on the 'Hartley' variety (Schaad and Wilson 1971). Cultural practices which improve soil structure and drainage can induce resistance to this disease and restore vigor (Schroth *et al.* 1977).

Late-leafing varieties such as 'Franquette' and 'Hartley' are less susceptible to walnut blight since their bloom occurs in most years after the spring rains. Time of leafing is, therefore, a good indicator for blight susceptibility (Table 3).

Early harvest period, though correlated with early leafing and blight susceptibility, can be important with regard to mold problems on mature nuts. Early-maturing varieties can be harvested before the fall rains which aggravate mold infections. Again, this is typical ecological resistance since the susceptible stage, the mature nuts, and the environmental conditions for infection are not in synchrony.

Observations over many years have shown that early varieties such as 'Payne', 'Ashley' and 'Serr', develop high codling moth infestations while 'Hartley', 'Franquette' or 'Mayette', late varieties, are seldom severely infested. Nut growth in early varieties is better synchronized with the phenology of this pest. Consequently, a full first generation can develop on these varieties which will eventually lead to high seasonal infestation levels. On late varieties nut growth is delayed, few suitable oviposition sites are available during spring emergence and thus only a small first generation will develop. Seasonal infestation levels will also be considerably lower on these varieties.

Good shell seal is an important varietal trait for the handling and processing of nuts. Closure of the suture at the base of the nut may be important with respect to codling moth susceptibility. This characteristic would also affect the ability of other pests such as navel orangeworm and the filbertworm to enter the nut.

Differences in varietal susceptibility to walnut husk fly attack were noticed soon after this pest became established in California. These differences were believed to be related to husk hardness since female flies are less able to insert their ovipositor in a hard husk. 'Franquette', 'Mayette', 'Eureka' and 'Payne' are susceptible, whereas 'Placentia', a variety with a harder husk, is somewhat less so (Boyce 1934). Studies of husk hardness as an indicator for husk fly susceptibility have not been conducted with any of the newer varieties. The effect of husk infestation on kernel quality and yield is more pronounced on late varieties since fly infestations begin when the nut is partially mature. On these varieties husk fly larvae cannot only cause stained shells and discolored meats but also moldy and shrivelled kernels. Nut maturation of early varieties is more advanced by the time husk fly begins to oviposit and thus damage to the kernel is not so severe.

Few observations are available on varietal susceptibility to leaf and bark feeding pest arthropods. Varieties differ in their bearing habit (Table 3) and in the amount of foliage supporting the nut crop. It would appear that the light-bearing varieties with a high foliage to nut ratio can tolerate more leaf-feeding or reduction in photosynthetic surface before the crop is affected than the heavy-bearing varieties. Indirect pests seem to have a greater impact on nut production and quality when they reach their peak densities during the active growth period of the nuts (spring and early summer). Predominantly mid-summer aphid populations had no significant effect on yield the first year (Barnes and Moffitt 1978). The second year of mid-summer and fall infestation reduced yield and also suppressed the number of staminate flowers the next spring by 80%. Mid-summer infestations of European red mite had no effect the first or second year on yield and quality but resulted in severe, apparently cumulative damage (40% depression in yield) at the end of the third year. Severe depression of staminate flowers was also observed the third year (Barnes and Moffitt 1978). The 'Eureka' variety is less susceptible to the development of mite populations than interplanted 'Payne' or 'Placentia' (Barnes 1978).

Regional Differences in the Pest Complex

The five major growing regions are climatically very distinct as indicated by the normal rainfall and average July temperature in Table 2. These regions differ also in varietal make-up, crop associations, native vegetation and cultural practices. Many walnut pests have relatively specific ecological requirements which determine their regional distribution.

Blackline has been more frequent along the central coast district, area 3, (Fig. 1), but is now on the increase in the Sacramento Valley and the northern part of the San Joaquin Valley (Serr 1959). Deep bark canker on the other hand is favored by high temperatures during summer. Presumably for this reason, this disease is more common in the warm interior valleys, primarily the San Joaquin Valley (Schaad and Wilson 1971). Walnut blight is more severe in the humid coastal districts and in parts of the Sacramento Valley with spring rainfall, since moisture is a prerequisite for infection.

The codling moth is present in all walnut growing areas of California. The walnut husk fly is widely distributed and equally damaging in most areas, though at present it has not been reported from commercial orchards in the southern San Joaquin Valley. The walnut aphid appears to be present throughout the growing areas. The dusky-veined aphid formerly appeared to be restricted primarily to the coastal areas. However, following the suppression of the walnut aphid by *Trioxys*, the dusky-veined aphid has developed damaging populations in the central valley.

Plant-feeding mites are also widespread, however, species composition and relative importance of each species vary regionally. In the northern central valley the twospotted mite is the most prevalent species on walnuts while the Pacific mite is the more important species in southern California. European red mite is found in pest status in regions subject to coastal influence. The carmine spider mite occurs on walnuts only in interior valleys of southern California.

Little is known about distribution of the several scale insects which attack walnuts in the major growing areas. The redhumped caterpillar and the fall webworm prefer the warmer interior growing areas and are relatively rare elsewhere. Their importance on walnuts is linked to presence of other host trees in an area as a reservoir for infestations. Similarly, the Pacific flatheaded borer is primarily a walnut problem in the foothills where native trees support endemic populations of this beetle.

THE TOOLS AND METHODS FOR WALNUT PEST MANAGEMENT

The Components of a Pest Management System

Pest management must be a flexible and responsive activity since it operates within the constantly changing biological system of pests, natural enemies and the host tree, all of which are affected by the weather and man's cultural practices. The status of the pest-crop system and the effect of previous control measures on it need to be evaluated continuously before new control actions are undertaken. A pest management system must have certain essential components and be organized in such a way as to facilitate this continuous review. Haynes *et al.* (1973) and Tummala and Haynes (1977) give a clear description of the necessary components in a pest management system and how these components relate to one another.

The central component is a "predictive management model" which, based on climatic and biological data, provides recommendations for control actions. The management model may take many forms - a simple empirical model represented as a chart or a table, or a more complex biologically explanatory population model which requires a computer facility to access it. Whatever form this management model might have, it should permit a pest manager to integrate and interpret biological and weather data collected in the field and help him evaluate control actions already taken or to be taken. The "biological monitoring" component supplies information on the pests, natural enemies, and the host plant to initiate and up-date recommendations by the predictive management model. Since the whole pest-crop system is keyed to weather, climatic input must come from the "environmental monitoring" component. Agricultural weather networks can supply this information.

From biological and climatic input the management model provides specific recommendations. The action which follows may consist of one or a combination of several control strategies to reduce pest populations below an economic threshold. This approach to pest management has already been put into practice in several crop ecosystems. Croft (1975b) applied it to apple pest management and showed how one can effectively manage a complex of phytophagous and predatory mites on a real-time basis.

Present Status of Integrated Pest Management (IPM) System Components on Walnuts

In order to put a pest management program into operation, the component parts (biological and environmental monitoring, predictive management models, set of control tactics) must be developed interdependently. In the following paragraphs the authors examine the status of these components with regard to the walnut pest complex (described earlier). Figure 7 summarizes which components are available or implemented for each pest and which remain, at present, in the research phase.

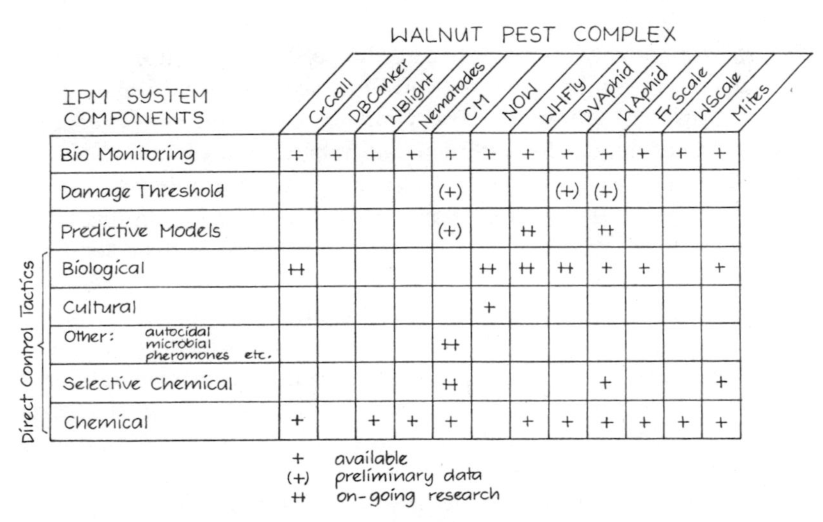

IPM SYSTEM COMPONENTS		CrGall	DBCanker	WBlight	Nematodes	CM	NOW	WHFly	DVAphid	WAphid	FrScale	WScale	Mites
Bio Monitoring		+	+	+	+	+	+	+	+	+	+	+	+
Damage Threshold						(+)			(+)	(+)			
Predictive Models						(+)		++		++			
Direct Control Tactics	Biological	++					++	++	++	+	+		+
	Cultural						+						
	Other: autocidal, microbial, pheromones etc.					++							
	Selective Chemical					++				+			+
	Chemical	+		+	+	+		+	+	+	+	+	+

WALNUT PEST COMPLEX

+ available
(+) preliminary data
++ on-going research

Figure 7. Status of integrated pest management (IPM) system components on walnuts.

Biological monitoring. This activity is pivotal to the pest management process. It establishes when, where, in what numbers, and in what stage a pest or a natural enemy is present. Methods of monitoring are available for most walnut pests, however, their accuracy for measuring pest density is either highly variable or is not known. Key walnut pests such as the codling moth and the husk fly are commonly monitored with attractant traps which are sensitive at low density and which can be operated with little effort. Ease of monitoring with these methods is contrasted with the difficulty in interpreting results in terms of larval activity and density. Many other pests on walnuts can be sampled directly by inspecting the leaf surface, twigs, or the bark. However, information on density must be sufficiently accurate since management decisions are directly dependent on what was observed in the field. In order to obtain population estimates of known precision, dispersion and sampling statistics must be known. Sampling procedures for optimization of sampling effort and accuracy need development.

Environmental monitoring. Detailed knowledge of the climatic history of an area is fundamental for walnut production. Although historical weather data are valuable to describe the local climate and its extremes, they are of limited use for crop management on a seasonal, real-time basis. Current weather information and forecasts of certain weather elements could be utilized in many ways by the orchard manager: for frost predictions in spring or fall, scheduling of irrigations, predictions of bloom, nut maturity and harvest time, and for timing of control applications with regard to insect phenology and disease epidemics. To be useful for pest management, weather information must be collected and reported daily or even more frequently depending on the data requirements of a predictive pest management model.

Only a weather network dedicated to the needs of agriculture can provide this information. Presently such a network does not exist in the walnut-growing regions of California although a few counties make limited use of current weather data and forecasts for *e.g.* timing of walnut blight sprays, scheduling of irrigations, and frost alerts. Recently a cooperative effort between NOAA (R.R. Stephen, Agricultural Meteorologist) and the University of California Cooperative Extension Service has begun to initiate a pilot network in the San Joaquin Valley (area 1).

Predictive management models. Except for the codling moth, phenological or density models for most walnut pests are lacking which could aid with decisions concerning timing or the need for specific control applications. For timing, phenological models are particularly suitable for key pests which cannot be monitored with precise methods and which require chemical controls in most years.

Damage thresholds. To determine the need for intervention with a specific control measure, it must be known how pest density and duration of feeding relate to crop loss. However, these relationships and their effects upon yield and quality are not known for most walnut pests.

Crop loss is easier to define for the direct pests: codling moth, navel orangeworm, and husk fly. For these pests grade levels are set according to industry standards. Internal damage to the kernel attributable to insects may not exceed 5% on U.S. No. 1 walnuts. This injury level applies to shelled and unshelled walnuts. In addition, on walnuts sold in-shell, not more than 5% of a lot may have shell stains caused by husk fly.

Leaf and bark feeding pests such as mites, aphids, and scales have a more subtle and indirect effect on yield and quality of nuts which is more difficult to observe and measure. The damage potential of high mite and walnut aphid populations has been established (Barnes *et al.* 1971; Barnes and Moffitt 1978). For the walnut aphid, populations above 10-15 aphids per leaflet for a 5-6 week period have been shown to result in severe loss (Barnes *et al.* 1971). Only for one indirect pest, the dusky-veined aphid, are more precise data available correlating density and yield reduction (Fig. 9).

Management methods for the pest complex on walnuts. Presence of walnut blight bacteria in dormant buds and catkins can be monitored with a selective medium. In areas with spring rains during bloom, a minimum of three protective sprays are usually recommended. The first one occurs at 25% catkin expansion, the second between 30-50% pistillate bloom and the third post bloom. This strategy has been shown to be more effective than multiple applications during pistillate bloom. Additional sprays may be necessary depending on frequency and amount of rainfall during this period. It may be possible to determine more exactly if and when protective sprays need to be applied based on monitoring of the blight organism, weather observations and forecasts. Such a system has reduced spray applications against fire blight, *Erwinia amylovora,* on pears (Thompson *et al.* 1977) and scab, *Venturia inaequalis,* on apple (Jones 1976).

Considerable effort has been devoted to codling moth research over the past hundred years (Butt 1975). Because of its economic importance, much research has concentrated on control and questions related to it. On walnuts, as on other deciduous fruit crops, the use of broad-spectrum chemicals for its control interferes with biological control of such pests as phytophagous mites, aphids and scales. A more favorable environment for natural enemies will result if a selective compound can be developed for codling moth control. Until recently bait pans and light traps were commonly used for monitoring, but these more laborious methods have been replaced by the pheromone trap which has been widely accepted (Batiste 1972; Madsen and Vakenti 1972). However, the major difficulty has been with the interpretation of catches. Several recent research papers established the pheromone trap as a valuable tool in apple orchards for damage estimation (Riedl and Croft 1974), and for determination of economic thresholds and timing of control measures (Madsen and Vakenti 1973; Riedl *et al.* 1976). In addition, physiological-time models with monitoring input from pheromone traps are being developed for phenological predictions and as additional aids for codling moth management (Falcon *et al.* 1975; Riedl and Croft 1978a, b). For a recent review of monitoring and forecasting methods for codling moth management, primarily on pome fruits, see Riedl (1979).

The codling moth on walnuts is a distinct host race with biological and behavioral characteristics different from populations on apple and plums (Phillips and Barnes 1975). In addition, on walnut larval establishment is less successful than on apple and pear. Therefore, the relationship between trap catch and damage needs to be defined for walnuts and predictive models would have to include the biological parameters characteristic of the walnut race before they can be used with confidence.

Unless infestations are very severe, chemical control of codling moth requires treatment against only one of two broods. Timing of spray applications against the first brood of the codling moth on 'Payne' walnuts, when required, has been successfully linked to nut size. It was found that sprays against this brood were most effective when nut diameter reached 3/8-1/2 inches (Michelbacher *et al.* 1950). The second brood spray is now timed by pheromone traps. By virtue of this more accurate timing for second brood, an opportunity arises for integration of chemical control of codling moth and biological control of walnut aphid as discussed in the section on chemical control.

On early varieties in the Tulare County area in the southern San Joaquin valley it was determined that there was no need to apply a spray if cumulative catch per trap did not exceed 50 moths at the time of peak spring brood emergence. Optimal spray timing was at the catch peak of the overwintered generation and one to four days after the peak catch of the summer brood (Culver 1978a, b). Whether this system of spray timing and the critical catch levels can also be extended to other walnut-growing areas needs to be verified.

Both sexes of the adult walnut husk fly respond readily to various baits. Three different bait-trap methods can be used for monitoring husk fly adults. The most effective, but least convenient method is a liquid bait consisting of glycine and lye (Boyce and Bartlett 1941a; Barnes and Osborn 1958). For many years, particularly during the time when husk fly began to expand into northern California, a sticky ice cream carton with ammonium carbonate as attractant has been employed. This trap, commonly known as the 'Frick' trap (Frick 1952) was originally developed for the cherry fruit fly, *Rhagoletis cingulata* (Loew), following the discovery by Hodson (1948) that ammonia-releasing substances were attractive to the apple maggot, *Rhagoletis pomonella* (Walsh). The third trapping method now in use is the commercially available Zoecon[1] AM (apple maggot) trap which has proven to be very effective for husk fly. It consists of yellow cardboard coated with a sticky substance containing protein hydrolysate and ammonium acetate as attractants. The ammonium carbonate trap as well as the AM trap give similar flight curves (Fig. 8a, b). The latter is, however, more efficient and catches a larger number of flies.

Interpretation of the catches provided by any of the husk fly traps is a serious problem. Researchers have not been successful in relating fly catches in these bait traps to oviposition as climatic and cultural factors affect the time when varieties become susceptible to oviposition by maturation and softening of the hull. Presently, the first treatment is recommended ten days after fly catches begin to show a steady increase (Joos *et al.* 1974). Patterns of catch and oviposition in two walnut orchards in northern California during 1977 showed little similarity while the schedule of emergence was similar at both sites. (Fig. 8a, b). The observed differences between orchards is probably related to the varieties which become susceptible at different times.

Monitoring for egg punctures and the beginning of larval development might be a better method to determine when to spray. The grade level for shell staining is presently set at 5% and directly correlated with larval infestation. This level or a more conservative figure could be used in a sequential sampling plan to determine spray need. Although monitoring of adults is more convenient and provides more lead time for treatment, direct monitoring of infestation will be more accurate if carried out according to an adequate sampling program.

There is a great need for a sophisticated management model for this fly. However, the biological data base for construction of such a model is not nearly so extensive as for the codling moth and much basic research needs to be done.

Ovipositional activity of the navel orangeworm has been directly monitored in almond orchards with an egg trap described by Rice *et al.* (1976). However, this trap has not been used in walnut orchards for timing purposes, since this pest can be effectively controlled by sanitation and harvest management.

A monitoring method for egg-laying would be preferable for many pests since it avoids the inferences which have to be made from adult catches to oviposition. No attempt has been made to determine a damage threshold based on eggs per trap and infestation levels. Research on the chemistry of the navel orangeworm female moth sex pheromone is underway by USDA scientists.

The filbert worm is sporadic and only of local importance but unsuspected infestations may arise which could be detected by monitoring. This insect

---

[1]Zoecon Co., Palo Alto, CA.

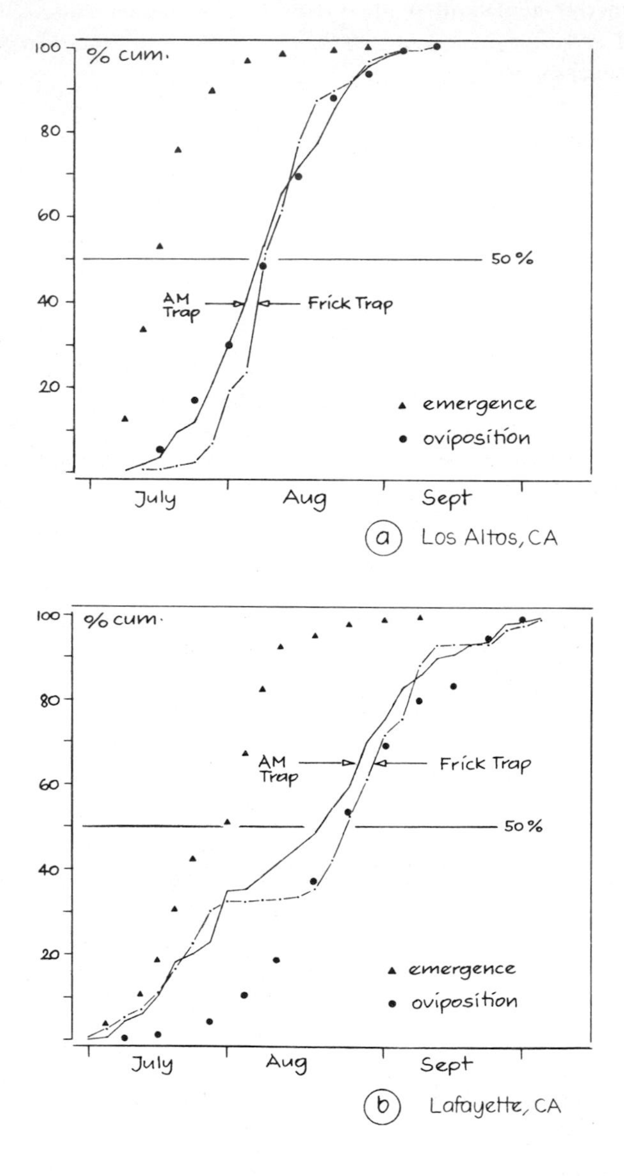

Figure 8. Seasonal phenology of the walnut husk fly at a) Los Altos, Calif. and b) Lafayette, Calif., during 1977 (H. Riedl, unpublished data).

responds to bait pans and has been monitored in the past with this method (Michelbacher *et al.* 1956). A pheromone trap would be more convenient and economical and thus permit routine surveys in areas where this pest has been known to occur.

The fruittree leafroller spills over to walnuts from other fruit crops. Its presence in walnut orchards can be easily detected with pheromone traps and verified with leaf sampling. Most other lepidopterous leaf feeders which have been reported from walnuts cause conspicuous damage and they can be detected through routine orchard inspection.

No definite statistically sound monitoring procedures are available for the walnut aphid and the dusky-veined aphid and their natural enemies. Commonly used sample sizes have ranged from 50-100 leaflets taken at random from throughout an orchard. The terminal leaflet is usually excluded from the sample. A threshold for the dusky-veined aphid could be derived from further data confirming relationships in Fig. 9 in which the proportion of large-sized nuts decreases as the percent leaflets with aphids increases (Olson 1974).

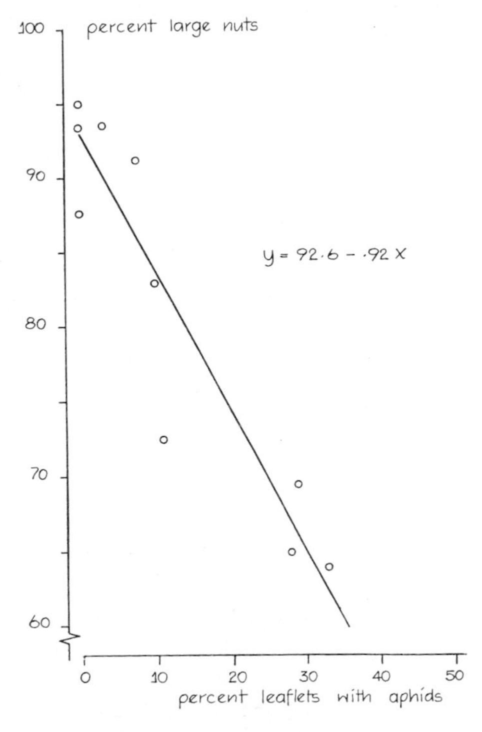

Figure 9. Dusky-veined aphid: relationship between percent leaves infested during spring and nut size at harvest (From Olson 1974).

Soft scales develop on the twigs during late winter but move as young crawlers to the leaves in spring. Armored scales on walnuts feed primarily on the bark. Population estimates can be obtained by sampling units of bark surface or, in the case of the summer populations of soft scales, by taking leaf samples. Relationships of density to crop loss still need to be investigated for all scales on walnuts and consequently no damage thresholds are available. Control measures are timed according to crawler emergence. Pheromone traps with virgin females have been used on an experimental basis for monitoring emergence of San Jose scale males. Efforts to identify the pheromone of two armored scales, the San Jose and the walnut scale, are in progress (Rice 1978).

The standard sampling scheme for mites on apple has been to collect 100 leaf samples, 10 each taken sequentially from the outside to the inside of 10 trees (Croft 1975c). However, a subsequent analysis indicated that this standard sample size was inadequate and that the precision of population estimates obtained with this scheme was poor (Croft *et al.* 1976). An improved sampling program for mites and their natural enemies on walnuts is certainly a prerequisite for their management and could form the basis for the construction of empirical decision-making indices for mite control on walnuts. Croft (1975c) described such an index for the mite complex on apple in Michigan. Although the detrimental effect of high mite densities on the nut crop has been demonstrated (Barnes and Moffitt 1978), there are no published guidelines which would tell a grower at which densities to treat.

Biological Control

The more important natural enemies of walnut pests were already discussed. Few fruit crops have as strong a biological control component as walnuts. The equilibrium of most indirect pests of this crop is well below the damage threshold. Therefore, the timing, choice of chemical and dosage must be carefully considered to minimize impact on the natural control forces in a walnut orchard. The search for additional biological control material for two key pests, the navel orangeworm and the walnut husk fly is being continued. The case history of an integrated control program for walnuts discussed under chemical control exemplifies how a spray program can be modified to accommodate biological control.

One area which is also critical to the success of biological control on walnuts in California but has received little attention is pesticide resistance in natural enemies. On apple in Washington, Michigan, and other states, biological mite control is only possible because several effective mite predators, primarily phytoseiid mites, have acquired resistance to azinphosmethyl and other chemicals which are frequently applied against the key pests (Hoyt 1969; Croft 1975c). Azinphosmethyl and other OP materials have been used for codling moth control on walnuts for the past 15 years. Although selection pressure is not as high as in other deciduous fruit crops due to lower pesticide use on walnuts, it is likely that various groups of predators and possibly also parasites have acquired a certain degree of resistance over this time period of continuous exposure. Knowledge of tolerance and resistance patterns in natural enemies of walnut pests would undoubtedly help evaluate or redesign control programs for key pests in order to augment the role of biological control agents.

In comparison with insect and mite pests, possibilities for biological control of plant disease organisms are limited. However, recent studies suggest that one serious disease of walnuts and other fruit trees, crown gall, may be controlled by biological means. In large-scale field experiments a non-disease producing strain of the crown gall bacterium, originally discovered in Australia (New and Kerr 1972), was able to protect seedlings of several stone fruit trees against the gall-forming strain (Moller and Schroth 1976). Tests with walnut seedlings gave similar results (Schroth 1978).

Cultural Control

Normal cultural practices in a walnut orchard such as soil management, irrigation, fertilization, pruning, etc., are primarily designed to promote optimal tree growth and nut production. These activities affect tree condition and the orchard environment, and thus they may influence disease and arthropod pest populations as well. Particularly, soil management practices seem to be relevant to the management of walnut pests. Experience in other deciduous fruits suggests that the ground cover is important for biological control of orchard pests. For instance, in Michigan apple orchards a small weeded area below each canopy is considered essential to inoculate each tree with predatory mites in the spring (Croft 1975c). As mentioned before, dust associated with clean-cultivation is generally implicated to induce mite build-up in walnuts. However, lack of refuge for mite predators in clean-cultivated walnut orchards may be an equally important factor. In a population study of the walnut aphid it was shown that the ground cover acted as an insectary for coccinellid predators of this aphid which contributed irregularly to suppression in the tree (Sluss 1968). A grower can choose from several soil management systems which maintain a cover crop of volunteer or sown plants during part of the season or throughout the year (see earlier section on soil management practices). The horticultural advantages and disadvantages of the various soil management systems are well known, but their benefits to pest management, particularly biological control are poorly documented. Questions which seem to be relevant in this regard and need to be investigated concern the species composition of the ground cover, time of year when ground cover needs to be maintained to support the natural enemy complex and what system optimally combines benefits to pest management and tree growth.

One pest, the walnut husk fly, could be directly affected by soil management practices since it pupates in the top few inches of the soil. The effects of various soil management practices on pupal survival need further investigation (Boyce 1934). To a certain extent the climate in an orchard is also influenced by the kind of soil management system being used. Through proper choice of cultivation, it may be possible, therefore, to make an orchard habitat less favorable for certain pests such as the temperature-sensitive dusky-veined aphid.

The method of irrigation should also be evaluated in the context of pest management. Irrigation water applied by sprinkler often wets the lower tree canopy which can aggravate walnut blight on susceptible varieties (Schroth and Mulrean 1977). Flood irrigation, particularly during the cooler post harvest period in the fall and winter, can lead to incidence of *Phytophtora* on susceptible root stocks, unless trees are planted on low ridges or berms not reached by irrigation water (Rackham and O'Reilly 1967). Orchard sections which receive a more than adequate water supply during the season reportedly produce heavy populations of walnut husk fly (Boyce 1934). Whether high soil moisture directly increases pupal survival in the soil or whether nuts become more suitable for husk fly attack under a high moisture regime needs further investigation.

Chemical Control

The development and use of pesticides in walnut orchards represents the accomplishments of many entomologists, plant pathologists, plant scientists, and chemists dealing with ecological and economic reality. When populations of arthropods, microorganisms and weeds interfere with economic plant-productivity, they are pests and intervention with pesticides is frequently required. That this may be done selectively depends primarily on the availability of pesticides with a varied spectrum of physiological action. The future success of integrated control on walnuts hinges to a great extent on the ability of the pesticide industry to responsibly develop and maintain a choice of selective chemicals, under the scrutiny of those charged with the protection of the environment and food from unwarranted pesticide residues.

The pattern of development of chemical control in walnut orchards in California reflects a process of maturation of an irrigated agriculture over a period of a century. As has been noted, this has been accompanied by a shift in locale of orchard plantings, adaptation of the codling moth to walnuts, invasion by three important pest species (walnut husk fly, navel orangeworm and dusky-veined aphid), and introduction of one effective parasite (for the walnut aphid). These changes have been accompanied by our increasing ability to control pest species in a partially integrated fashion through the development and selection of an appropriate variety of organic and inorganic pesticides and plant growth regulators. Greater tree density and smaller trees have eased the problem of adequate coverage which once necessitated use of cumbersome spray-towers in earlier days of walnut orchard pest control.

The inorganic-botanic pesticide period, 1918-1946. No chemical protection was available against pests in walnut orchards until about 1918. Prior to 1900, walnut growers suffering heavy losses from walnut blight offered a reward of $20,000 for a remedy (Smith 1953). The walnut aphid also caused severe losses annually. Chemical control measures against the codling moth were one of the earliest to be introduced. Following the development in 1918 of significant codling moth infestations in walnuts in Orange County in southern California, Quayle (1926) developed the use of a single first brood spray of basic lead arsenate. This program remained essentially unchanged until the introduction of DDT, modified only by use of spreading agents and improvements in application equipment. As the walnut industry developed in the central valley, it was found that two sprays were required against the more severe infestations which eventually developed there. Acid lead arsenate plus basic zinc sulfate safener to prevent foliage injury was more effective and preferred in the north (Michelbacher *et al.* 1944). Acid lead arsenate could not be used on the predominant varieties in the south because of such injury. After 1932, spraying was done by hand from wooden masts and hydraulically operated steel towers 30 feet high to accomplish the thorough coverage required with arsenicals (Boyce and Bartlett 1941b). These were gradually superseded by vertical booms with oscillating guns (Fig. 10A).

Because of the awareness of growers and advisors of the problems created by DDT, basic lead arsenate was used against the codling moth in the south in Santa Barbara County until 1958. Its ultimate replacement in that locale was related not only to the improved control with DDT, but also to the complicating problems of the introduction of the navel orangeworm and the spread of walnut husk fly, the latter requiring broad spectrum organophosphates. The use of arsenicals persisted in northern California orchards for many years after the availability of DDT.

Among the early contributions of walnut pest management research (which extended beyond the walnut orchard) was the origination of the use of nicotine sulfate in dust formulations as an insecticide. Professor R. E. Smith, Division of Plant Pathology, University of California, Berkeley, developed this formulation for the control of walnut aphid in connection with his research on walnut blight, as he suspected aphids to be vectors (Smith 1921). Thereafter entomologists and growers worked with nicotine dusts on many crops for over 30 years in many parts of the world. The first commercial manufacture of nicotine dusts was by a cooperative walnut growers' association in 1918 (Campbell 1923). From that

Figure 10. (A) Vertical boom sprayer with oscillating spray guns in operation. This type of sprayer was commonly used during the 1940's.
(B) Air-carrier sprayer: note special chute to direct spray into the upper canopy.

time until the early 1950's, nicotine sulfate-lime dust formulations were an accepted product for walnut aphid control, along with nicotine sulfate sprays. The use of dusts predominated during much of this period, and successful treatment depended upon ideal temperature and lack of wind, as well as proper formulation. No insecticide resistance was observed; however, the use of nicotine dust treatments for walnut aphid control would result in resurgence of aphid populations, if mistimed in relation to the occasional presence of adequate predation (Michelbacher and Swanson 1945). This effect resulted from the virtual elimination of prey by effective dust treatments, much as routinely occurred in the initial use of organophosphates. Dust deposits from repeated applications for walnut aphid also resulted in serious spider mite infestations (Michelbacher and Bacon 1952a).

Walnut blight has been the subject of research since 1901 and annually has caused very severe losses. Early efforts to control this disease with bactericides (reviewed by Fawcett and Batchelor 1920) were unsuccessful. Finally, in 1927 B. A. Rudolph began his work which demonstrated that the key to success was proper timing and the use of 1 to 3 applications of Bordeaux mixture (Rudolph 1933). With downward adjustment in dosage to avoid severe foliage injury, and with refinements in timing, this recommendation still stands, along with the use of fixed copper compounds (Teviotdale and Moller 1977). From its first use, Bordeaux mixture was observed to induce aphid populations, presumably by suppression of partially effective predation.

Following the introduction of the walnut husk fly into southern California in the mid 1920's, a comprehensive investigation of its biology, ecology, and control was undertaken in 1928 (Boyce 1934). Control measures using cryolite dusts and sprays were developed by 1932 but the degree of control resulting from their use was only fair by present standards (Boyce 1932, 1936). Cryolite presisted in recommendations for walnut husk fly control until 1956 (Ortega 1956b).

Toward the close of the inorganic insecticide period, another notable achievement in walnut orchards with broad application in pest management was the development of the first synthetic organic acaricide, dinitro-o-cyclohexylphenol or DNOCHP (Boyce *et al.* 1939). This compound replaced sulfur dust applications that often caused severe injury to walnut foliage. DNOCHP was used as a 1% dust (DN Dust) and large-volume, low-velocity dusting machines were developed for its application (Boyce and Bartlett 1939). The dicyclohexylamine salt of DNOCHP (DN-111) became the first widely used acaricide in deciduous orchard sprays.

In the decade prior to 1946, programs for pest management in walnut orchards of varieties susceptible to codling moth consisted of (a) 1 or 2 treatments of Bordeaux for blight, (b) 1 or 2 sprays of lead arsenate for codling moth (plus nicotine sulfate if aphids were present) followed by (c) 1 or 2 nicotine sulfate dusts as required for aphids, (d) a cryolite treatment for husk fly (portions of southern California only) and (e) a DNOCHP dust treatment if mites developed, – 3 to 5 treatments per season, according to locale and intensity of the blight and aphid problems. Weed control was by disking 3 to 5 times a year. The Bordeaux treatments often caused plant injury and tended to induce aphid populations, nicotine dusts induced mite populations, the control of codling moth and husk fly was less than satisfactory, and the lead arsenates gradually accumulated in the soil.

Organic pesticides – the path to integration, 1946-1970. The availability of a broad range of pesticidal compounds with varied physiological action is a prerequisite for the selection of chemicals which due to their specificity, fit into the management of a particular agricultural ecosystem. Those which have an efficacious spectrum of action, are not prone to resistance by the pest complex, and are minimally damaging to biotic regulation, may eventually be chosen for use in integrated programs to avoid ecological and biological reaction. Such upsets may take place rapidly and be readily determined, *e.g.* suppression of beneficials and resurgence of pests, or may take many years to appear, *e.g.* the development of resistance, which, based upon the experience in the walnut orchard, may or may not take place.

The development and use of organic pesticides in walnut orchards provides a case history paralleled by other orchard experience. It was accompanied by a search for integration of chemical, biological, and cultural control. Indeed, the first published use of the term "integrated control" appears in a publication of Michelbacher and Bacon (1952a). DDT, in a single spray against first brood, provided distinctly superior control of codling moth as compared with the arsenicals, but its introduction was followed by secondary problems not to be resolved in a satisfactory manner until the development of other insecticides and the integrated programs discussed later. Populations of the walnut aphid were enhanced and those of tetranychid mites, and unarmored scale insects rose to pest status in many orchards when adequate dosages of DDT for codling moth were used (Michelbacher and Ortega 1958). However, DDT replaced basic lead arsenate rapidly in most of southern California, though it was not initially recommended unless nicotine was available for aphid control (Boyce 1946). The navel orangeworm invaded California and became prominent as a serious pest at hullsplit of new crop walnuts in Riverside County in 1946. It was demonstrated that the suppression of this scavenger was accomplished best if DDT were used for codling moth control (Ortega 1950b), and if trash walnuts were destroyed by disking (Michelbacher and Ortega 1958). This insect again became important when DDT was lost.

Initially, no completely satisfactory control was available for the serious populations of mites and frosted scales which sometimes followed DDT trials for codling moth (Michelbacher *et al.* 1947; Ortega 1948). Because of these secondary outbreaks, DDT alone was not recommended at full dosage in northern California (Michelbacher and Middlekauff 1948). Standard lead arsenate plus basic zinc sulfate safener was preferred in the north, while basic lead arsenate continued as an option against moth populations in the south (Michelbacher *et al.* 1950; Ortega 1950a). However, Michelbacher and Swanson (1946) had found that a low dosage of DDT was not followed by serious mite populations in northern orchards but was inadequate for codling moth control. The addition of a low dosage of DDT to standard lead arsenate improved control. This continued as a recommendation as late as 1953, but only when high pressure, full coverage sprayers were used, which were then being rapidly replaced. Another factor which hastened the adoption of DDT was the introduction of labor-saving, air-carrier sprayers (Fig. 10B). These did not provide the thorough spray coverage necessary for success with lead arsenate. Thorough coverage was not necessary with DDT as this insecticide provided residual kill of moths, suppressing oviposition, as well as providing contact larvicidal action.

Nicotine was gradually substituted for aphid control by tepp and BHC (the latter was supplanted by lindane because of flavor effects), and by parathion. Nicotine continued to be recommended, when available, because of its selectivity, usually favoring aphid predators (Michelbacher and Middlekauff 1949).

The excellent frosted scale parasite, *Metaphycus californicus* (Howard) was very susceptible to applications of DDT against first brood codling moth. In fact, the action of DDT led to the awareness of the importance of this parasite. Applications of parathion and other organophosphates for aphid control effected excellent control of 1st and 2nd instar frosted scales present during spring and summer, though some time passed before this side benefit was recognized (Michelbacher and Swift 1954).

It should be noted that destructive mite populations develop in walnut orchards in some seasons though undisturbed by insecticides (Middlekauff and Michelbacher 1951). The introduction of Aramite for mite control, and the availability of parathion and malathion, provided seasonal control of mites and aphids in a single treatment in combination with the DDT spray for codling moth (Middlekauff and Michelbacher 1951; Michelbacher 1953). At first, some organophosphates used for aphids, including schradan and demeton, controlled mites as well (Michelbacher 1954). In most orchards, a single treatment, *e.g.* DDT plus schradan, gave seasonal control of codling moth, navel orangeworm, aphids, frosted scale, and mites (Michelbacher 1953; Ortega 1956a). Though such relief proved to be temporary, it was effective and economical for several years. Although schradan tended to increase calico scale populations to a noticeable degree, it did not interfere with subsequent heavy predation by Audubon's warbler or the rapid redevelopment of populations of the effective parasite, *Blastothrix longipennis* Howard (Michelbacher 1956). Demeton tended to increase frosted scale, European fruit lecanium and calico scale (Michelbacher 1957), but parathion, carbophenothion or ethion (the latter also acaricidal) readily controlled these scale insects when used in dormant sprays or if used for aphids or mites in summer.

Parathion resistance by the walnut aphid was first noted in San Jose (area 3; Fig. 1) in 1952 but did not develop at Linden (area 1) until 1958. In successive seasons it became evident that organophosphate programs against aphids were less and less effective. Populations highly resistant to malathion were controlled by parathion. Those tolerant of parathion were initially susceptible to schradan and carbophenothion. After continuous use for 6 years, schradan resistance appeared. All of the OP resistant strains could be controlled by lindane or endosulfan (Barnes *et al.* 1961). The systemic organophosphate, phosphamidon, was introduced for control of aphids as well as codling moth (Madsen *et al.* 1964) and, supplemented with an acaricide and a treatment for unarmored scales, enjoyed an extended period of use, involving no more than 2 sprays per season for insects and mites.

After 3 to 5 years of use, organophosphate resistance developed in local populations of the European red mite, twospotted spider mite and Pacific spider mite, as well as local resistance to ovex, dicofol and tetradifon. Aramite (discontinued in 1968) and its successor, propargite, which is structurally related and presumably similar in mode of acaricidal action, have not induced resistance in mite populations, though in successive use for the past 26 years.

By 1969, organophosphate resistance in aphid populations, specifically to parathion, malathion, diazinon, carbophenothion, demeton, schradan and ethion, was relatively widespread. Nicotine sulfate was available, endosulfan resistance was suspected in local areas, but phosphamidon remained generally effective. Phosphamidon's residual action gradually became limited, and 2 or 3 applications eventually became necessary, though the subsequent applications served as well for codling moth and husk fly where present (Barnes *et al.* 1969). The capacity of this damaging, monophagous aphid to develop insecticide resistance began to stress our ability to deal with it by use of synthetic organic insecticides. This set the scene for the spectular relief afforded by the introduction of the walnut aphid parasite, *Trioxys pallidus* (Haliday) (van den Bosch *et al.* 1970). The programs developed to take advantage of the control potential of this parasite by adjustment of the codling moth program are discussed under the section on integrated pest management.

In the 1960's, azinphosmethyl (Barnes and Ortega 1958) replaced DDT for codling moth control in many orchards. Similar to DDT, a single application was effective if directed against the first brood, when the nuts of the 'Payne' variety had a cross-section of 1/2-inch diameter (Michelbacher and Ortega 1958). The control of codling moth attained was exceptionally high, unarmored scales were controlled as well, aphids were often temporarily controlled, however, suppression of navel orangeworm was not so pronounced as with DDT (Bruce and Barnes 1973). The registration of DDT was cancelled in 1970. The approach to improved suppression of the damage caused by navel orangeworm following the loss of DDT is discussed under integration.

One phenomenon which developed following the change in spraying patterns in the mid to late 1960's was a general increase in populations of the walnut scale, listed as "not a serious pest" by Michelbacher and Ortega (1958). This seems related to altered spray practice, although the specific causative element is not known. Possibly the predominant and extensive use of phosphamidon during those years for aphids, codling moth, and walnut husk fly control caused the rise in walnut scale. This compound lacks action against scale insects. Walnut scale may be controlled by spraying with methidathion at leaf-bud swell (Sibbett and Davis 1978).

The walnut husk fly feeds on foliage surfaces, apparently ingesting honeydew. This habit makes it highly susceptible to residues of organophosphates. Sprays of malathion, parathion, ethion or carbophenothion are effective against the adults and are usually applied in August, timed by traps or by the onset of oviposition. Bait sprays of corn protein hydrolysate plus malathion or other OP's, applied in low gallonage by ground or aircraft equipment, are also highly effective if applied at onset of oviposition (Barnes and Ortega 1959; Barnes *et al.* 1969). Phosphamidon is also useful, especially after oviposition and hatching are well underway due to its systemic properties (Nickel and Wong 1966). It is often necessary to include acaricides, *e.g.* propargite or cyhexatin, in husk fly sprays, since organophosphate sprays in mid-summer may induce mite populations.

After harvest, hulling and drying, the entire walnut crop is fumigated with methyl bromide, primarily to suppress further development of navel orangeworm larvae. This procedure was developed in the research of Gerhardt *et al.* (1951) and also controls storage pests (Table 5) which may have established.

Nematodes were recognized as serious pests of walnuts in 1923 (Hodgson 1923), but no treatment was available until DD®, a mixture of 1,3-dichloropropene and 1,2-dichloropropane, was developed as a pre-plant treatment against root-lesion nematode by Allen (1952). On established trees, repeated applications of DBCP, 1,2-dibromo-3-chlorpropane (now suspended) gave fair results (Lownsbery *et al.* 1969).

Soil fumigation with carbon disulfide was the only pre-plant treatment available for *Armillaria* root fungus until methyl bromide was introduced for this purpose (LaRue *et al.* 1962).

Vegetation management (weed control) in walnut orchards is important to achieve maximum growth and production (Elmore and Lange 1977). The advantages and disadvantages of common soil management systems in walnut orchards were discussed in previous sections. The use of herbicides in complete chemical control or combinations of sod culture with strip chemical weed control under the trees has become rather common (Osgood 1977). Diuron and simazine were introduced by Day *et al.* (1964) and this was followed by elaboration of herbicide use in walnut orchards by A. H. Lange and others (see Table 7). When vegetation management with herbicides is practiced, certain pre-emergence herbicides are used for annual weeds followed by contact herbicides in spot treatments of perennial weeds or weeds tolerant of the pre-emergence treatment. A list of herbicides used in walnut orchards with remarks on their effectiveness and limitations is presented in Table 7. For a discussion of weeds in fruit crops (including walnuts) and their identification, the reader is referred to Lange (1968) and Fischer *et al.* (1976).

During the developmental era of organic pesticides prior to 1970, walnut pest management programs on varieties susceptible to codling moth, consisted variously of (a) 1-2 sprays during the bloom and post-bloom periods of Bordeaux or a fixed copper compound for bacterial blight, (b) a single treatment of azinphosmethyl or DDT against first brood codling moth to which was added an aphicide, typically phosphamidon, and an acaricide, typically dicofol, (c) an additional treatment or two for aphids, and (d) in infested areas or orchards, an organophosphate spray in summer for control of walnut husk fly or unarmored scales – 2 to 4 sprays annually, according to locale and intensity of the blight and aphid problems. Control was generally very good, though organophosphate resistance by aphids was troublesome and DDT and its metabolites accumulated in the orchard soil.

Integrated pest management, 1970-date. Awareness of the importance of the ecological strand of pest management in walnut orchards was strong from the beginning of the modern pesticide era (Michelbacher 1945). An extended period of experimentation by researchers and use by growers followed which led to the necessary information on the direct and indirect effects of the new insecticides. The successive introduction of an array of new compounds greatly enhanced the ability to suppress insect pest damage, earlier a source of considerable frustration.

At first, following the introduction of organic insecticides, integration was a goal which was scarcely achieved. Fewer applications (1-3) were initially required for insect and mite control, but this was often based upon thorough,

Table 7.
Herbicides, registration status
and remarks on their use for walnuts.[a]

Trade name	Common name	Registration status	Use
PREEMERGENCE			
Karmex®	(diuron)	Orchards established 1 year or more	Good control of a broad spectrum of annual weeds. Does not control perennials. May split application between fall and spring.
Princep®	(simazine)	Orchards established 1 year or more	Controls many broadleaf and grass weeds, somewhat weak on barnyardgrass (watergrass), crabgrass and low growing pigweeds. Does not control perennials.
Devrinol®	(napropamide)	Bearing and non-bearing	Excellent control of grass from seed (annual and perennials). Often used as combination with simazine or diuron. Very safe herbicide if kept out of root zone of trees. Needs rain within 2 weeks after application.
Surflan®	(oryzalin)	Non-bearing trees only (label submitted for bearing)	Excellent grass control. Has looked best when combined with simazine or diuron. Does not control mustard or clovers well. Needs rain within about 21 days after application.
Casoron®	(dichlobenil)	Bearing and non-bearing	Requires soil incorporation for maximum effectiveness. Controls nutsedge (nutgrass) and many annual weeds.
Eptam®	(EPTC)	Established trees	Short residual-applied in the irrigation water. Excellent grass control and generally good on purslane. Does not control perennials.
Premerg®	(dinoseb)	Bearing and non-bearing	Prior to weed germination. Short residual.
Weed oil	(various names)	Bearing and non-bearing	Broad spectrum weed control. No residual. Injurious when sprayed on trunks.
Paraquat®	(paraquat)	Bearing and non-bearing	Broad spectrum contact. Often combined with simazine or other preemergence herbicides.
Ansar 529® and others	(MSMA)	Non-bearing only	Directed sprays for johnsongrass control. Does not control bermudagrass.
Roundup®	(glyphosate)	Non-bearing only	Directed spray for all annual and perennial weeds. Do not allow spray drift to contact walnut foliage.

[a]Taken from: Elmore and Lange (1977).

chemical suppresssion, not integration, *e.g.* "Aphid control with such insecticides as tepp, BHC, or parathion cannot be considered satisfactory unless the population is all but eliminated from the orchard" (Michelbacher and Bacon 1952a). The same authors, however, carefully adjusted DDT dosages for codling moth below the level of adverse effects upon spider mites and frosted scale, a clear approach to integration. This was possible only by adding a low dosage of DDT to lead arsenate and the use of vertical-boom, full-coverage spraying.

With the gradual and localized development of resistance in aphids and mites came resurgence of their populations. The spectrum of compounds which successively became available temporarily overcame this, often for a period of years. Satisfactory integration was not feasible under these conditions. Repeated applications were eventually necessary, finally resulting in the change to yet another compound.

As with any other crop, there were no quick solutions or easy victories in the development and practice of integrated pest management in walnut orchards. There is ample awareness that although integrated programs have now greatly stabilized insect pest management in walnut orchards, integration has yet been but partially achieved, and there must be constant watch for biological reaction.

Five developments made possible the present stage of integration of chemical, biological and cultural management of walnut pests: (1) the introduction of the walnut aphid parasite in 1968-69 (van den Bosch *et al.* 1970), (2) the development in 1969-71 of the use of the synthetic codling moth pheromone for accurate and selective timing of a spray directed against the second brood of codling moth (Barnes and Davis 1972), (3) the development of phosalone, an organophosphate with a useful and selective spectrum, – codling moth, dusky-veined aphid, and unarmored scales and favoring the aphid parasite (Davis *et al.* 1972), (4) the development of ethephon, a plant growth regulator releasing ethylene which provides for rapid and complete dehiscence and early harvest, avoiding navel orangeworm infestation (Olson *et al.* 1975), and (5) the standing availability of propargite and cyhexatin as effective acaricides, of azinphosmethyl for serious codling moth infestations, and of fixed copper bactericides for walnut blight. How these elements were interwoven into an integrated pest management program will be related in the following paragraphs.

Though of supplementary assistance, satisfactory biological control of the walnut aphid from native predatory insects such as convergent lady beetles and lacewings had clearly failed. Following the introduction and establishment of the Iranian strain of the parasite, *Trioxys pallidus,* in 1968-69, a classical demonstration ensued of effective insect pest suppression by a climatically adapted strain of a parasite from the geographical area of the evolution of the pest species. Where populations of 85 aphids per leaflet had been encountered, despite predation, levels were dramatically lowered to less than 1 per leaflet (van den Bosch *et al.* 1970). However, the spring codling moth spray of azinphosmethyl suppresses the parasite and seldom controlled the aphids, because of organophosphate resistance. The suppression of the aphid by the parasite noted in 1971 offered such spectacular relief in orchard varieties not susceptible to, hence not sprayed for codling moth, and was of such fascination to growers, that codling moth control was neglected in 1972. There followed a spectacular growth of navel orangeworm populations (which develop on codling moth injured nuts as well as other substrates) and larval damage was the highest ever experienced by the industry, averaging 13% statewide (Anon. 1973a). Dusky-veined aphids, formerly suppressed by competition from the treatments for walnut aphids became a problem in many orchards.

A single treatment of azinphosmethyl, the replacement for DDT, against first brood codling moth was generally in use in the period 1968-72, and continues to be very effective. Codling moth populations are so greatly suppressed by this treatment that no treatment for 2nd brood is required. Walnut growers thus control every other brood. However, as previously noted, azinphosmethyl in spring suppresses the activity of the aphid parasite (Davis *et al.* 1972). The walnut aphid is very damaging in the spring months, during the period of rapid growth in size of the walnut (Barnes *et al.* 1971). To achieve selective timing and to avoid interference with parasitism of spring populations of aphids, the first step in integration was to direct the codling moth treatment against the second brood instead of the first (Barnes and Davis 1972). Interference with biological control of the aphid is generally not a problem at this later time because aphid populations have usually been strongly suppressed by high temperatures. With the development of the codling moth pheromone trap, accurate and convenient timing of the second brood treatment in late June or early July was made possible. This timing may vary from year to year by as much as 24 days in the same locality.

Azinphosmethyl is still preferred for second brood in severe codling moth populations. It is supplemented by propargite or cyhexatin, if mite populations are induced. Another option against second brood codling moth is the emulsifiable formulation of phosalone. This insecticide (a) suppresses moderate infestations of codling moth, (b) controls susceptible mites, (c) thoroughly suppresses walnut aphid, if not suppressed by high temperatures or by parasitism (which lags at high temperatures short of aphid suppression), (d) permits re-establishment of aphid parasites in late summer and fall, and (e) controls dusky-veined aphid, frosted scale, and fall webworm. If codling moth populations are severe, two applications of phosalone, at an interval of 21 days, are required.

Each of these regimes of azinphosmethyl and phosalone against second brood codling moth greatly suppress aphid parasite populations, the former by an extended residual toxicity, the latter by eliminating aphids. However, *Trioxys pallidus* is highly dispersive and orchards of varieties which do not require treatment for codling moth, *e.g.* 'Franquette', serve as reservoirs from which parasites annually and effectively recolonize phosalone-treated orchards. Recolonization by this parasite accompanies re-establishment of low aphid populations from alates in late summer and fall after phosalone toxicity wanes.

The suppression of populations and damage from the scavenger insect, the navel orangeworm, is achieved by indirect methods involving (a) destruction by flailing or disking in spring of trash nuts missed by harvesters and fallen to the ground in winter; (b) control of walnut blight and codling moth; and (c) by use of ethephon to effect early and relatively complete mechanical harvest (Olson *et al.* 1975).

The growth regulator effect of ethephon applied to walnut trees was investigated initially by G. C. Martin in 1968 as an aid to harvest. Ethephon releases ethylene which promotes rapid hullsplit, making possible an early harvest, and preserving higher walnut quality. A shortened interval (7 days) between hullsplit and harvest, which accompanies use of ethephon, avoids damage by the navel orangeworm. This practice prevents significant navel orangeworm damage even though all other preventive treatments are neglected. Ethylene gas was used in the 1930's as a fumigant for loosening hulls on harvested, green "sticktight" walnuts. The discovery of ethylene as a plant growth regulator was a product of applied science. Observations made around 1905 indicated that oranges developed better color in packing houses with kerosene heaters (which produced minute quantities of ethylene) (Chace 1935). Efficient mechanical shakers, providing rapid harvest which made the use of ethephon practicable, arose during the second World War (Serr and Fairbank 1944).

The control of walnut husk fly in infested districts remains partially integrated with the use of organophosphates as before.

In orchards of the 'Franquette' variety which are resistant to codling moth and navel orangeworm, the walnut aphid and scale insects are usually suppressed by parasites. Mite problems arising in some years are controlled by propargite or cyhexatin, which favor predator survival. The husk fly and dusky-veined aphid remain an annual problem on these varieties in infested areas and are controlled by organophosphates.

Partially integrated pest management programs include (a) 1-3 applications of a fixed copper bactericide, typically copper hydroxide, for blight, (b) one spray of phosalone in late June or early July against second brood codling moth and also controlling unarmored scales and dusky-veined aphid, and combined with propargite for mites if the latter are organophosphate resistant, (c) an organophosphate in August for walnut husk fly in certain areas, and (d) a spray of ethephon to hasten harvest and avoid navel orangeworm – 3 to 5 annual interventions. Control is excellent.

Weed control may consist of 3-5 cultivations, or of a pre-emergence herbicide, *e.g.* simazine, combined with or followed by use of paraquat as a contact herbicide.

Integrated pest management is now implemented to a degree by professional pest control advisors (Culver 1978a, b) but full industry-wide implementation is still lacking at present. Contributions toward the further development of integrated pest management for walnut insect and mite pests may include (a) the development of an insecticide specific for codling moth in the walnut orchard ecosystem, with dimilin as a good prospect, if otherwise environmentally safe, (b) establishment in California of walnut husk fly parasites long known to be present in Texas and New Mexico, (c) fruitful search in Texas and Latin America for parasites of the navel orangeworm, (d) establishment of successful parasitization of the dusky-veined aphid, and (e) elucidation of the role of specific cover vegetation in walnut orchards as a shelter or reservoir of beneficial species, should any benefit therefrom be more than the cost of maintenance *e.g.* more water and nitrogen fertilizer.

Alternative Tactics

During the pre-pesticide era deciduous fruit growers had few effective means available to them to keep losses from insects and diseases at acceptable levels. Based on experience and an often intuitive understanding of pest ecology, control programs were attempted which often combined several tactics and can be considered forerunners of our modern integrated programs, though without insecticides. Codling moth control on apple, as it was practiced before the turn of the century, may serve as a good illustration (Fitz and Fitz 1872). On many apple varieties worm-infested fruit drop prematurely to the ground. The fallen fruit was gathered and destroyed or hogs were allowed to consume the fruit in the orchard thus causing larval mortality, particularly in the first brood. Entrapping the larvae after they issue from the fruit was considered most effective. This was done by hanging old cloth in the tree crotches or by twisting a hayband around the trunk to serve as pupation site. To make this method even more effective, old bark scales were removed and the area below the tree was kept clean from dried-up vegetation and mulch in order to deprive mature larvae of possible cocooning sites. The spun-up larvae were then crushed or the bands burned. A special committee of the American Pomological Association in 1871 found that a trap consisting of three narrow boards fastened to the tree trunk was most effective in trapping codling moth larvae (Fitz and Fitz 1872). In addition to banding, second brood larvae which spun up in shipping barrels were destroyed by immersing the barrels in hot water or by burning them. The combined use of thorough sanitation, predation by a mammal, pupal habitat management, and mass trapping of mature larvae gave the best control available at the time, though certainly inadequate according to today's commercial standards. There is renewed interest today in these non-chemical, laborious methods. They are being used by a small but growing number of so-called 'organic' farmers in combination with botanical insecticides.

Of all the pesticides applied on walnuts during a season the broad-spectrum sprays against the codling moth, particularly against the first brood, are potentially the most disruptive. Although the impact of these sprays has somewhat lessened over the years due to selective timing, choice of material and probably also development of resistance in certain natural enemies (predatory mites) to organophosphate materials, the development of alternatives remains a high priority. These alternatives to the conventional pesticides must, however, fit well into present walnut production practices and above all be economical.

The success of the sterile insect technique (SIT) program against the screwworm fly, *Cochliomyia hominivorax* (Coquerel), in the southern United States (Baumhover *et al.* 1959) led to similar programs for other insect pests, among them the codling moth. Preliminary laboratory and field experiments on small acreages in British Columbia and in Washington demonstrated the technical feasibility of this method for codling moth control (Proverbs *et al.* 1966; Butt *et al.* 1973).

In a more recent field test in British Columbia a population reduction of 99.6% was achieved over a two-year period by combining area-wide sanitation and insecticide treatments followed by releases of radiation-sterilized males (Proverbs *et al.* 1977). The application of the sterility method requires an area-wide approach to codling moth control and isolation of a treatment area to prevent reinfestation. The codling moth is present throughout California and, in many areas of the state, several of its hosts are grown adjacent to each other. Therefore, only a well-coordinated, large-scale release program on several commodities could achieve the population reduction necessary for commercial fruit and nut production. Such a release program cannot be implemented in California without both government and fruit industry support since costs and technological requirements are prohibitive for the individual grower. During the early 1960's there was interest in a sterile-male release program for the navel orangeworm (Husseiny and Madsen 1964), but it was abondoned after some preliminary experiments.

Microbials for codling moth control have received considerable attention in the United States and overseas. Spore preparations of *Bacillus thuringiensis* Berliner (BT) are insecticidal against many lepidopterous pests. Field tests on apple indicated that codling moth is not very susceptible to this pathogen. Six cover sprays with a commercial BT formulation did not provide economic control (McEwen *et al.* 1960). The most promising pathogen is a granulosis virus which has performed well in recent field trials giving control comparable to broad-spectrum insecticides (Huber and Dickler 1977). However, unless problems with virus production, formulation, application method and safety are resolved there is little chance that a walnut grower will be able to use this pathogen for codling moth control in the near future.

Of all pesticide alternatives which are presently under development against codling moth, control methods utilizing the pheromone are closest to field implementation. The importance and use of the pheromone for monitoring has already been mentioned. The codling moth pheromone has been used for control purposes in two ways: (a) to remove the male moths in a population and (b) to disrupt the mating process.

Trapping of codling moth adults (both sexes) with bait or light traps has not given commercial control although a certain population reduction could be demonstrated (Yothers 1930; Collins and Machado 1937). Sex pheromone traps are more efficient, particularly at low density, but failed to keep infestations below acceptable economic levels in similar experiments (Proverbs *et al.* 1975; MacLellan 1976), unless the orchard was isolated (Madsen *et al.* 1976). In order to make mass trapping more effective additional investigations are necessary to improve trap design, optimize trap density and placement.

An early attempt to interfere with the normal oviposition behavior of the codling moth was the use of artificial light during the egg laying period around sunset (Herms 1932). This led to some decrease in larval infestation, but not to economic control.

The mating disruption technique, also referred to as "confusion method," interferes with the communication between sexes by permeating the air with high levels of pheromone. With this technique it is important that the pheromone is slowly released from many point sources within an orchard. After experimentation with various slow-release formulations, a microfibre system consisting of open capillary tubes which hold the pheromone has given the best results. These fibres when distributed by helicopter into canopies of pear trees maintained a sufficient pheromone concentration in the air to prevent mating and oviposition. Preliminary field tests have resulted in seasonal control of the codling moth. Much of this work is presently being conducted by USDA scientists at Yakima, Washington. The fibres containing the pheromone cannot be applied through conventional orchard sprayers. Specialized ground equipment needs to be developed so that this control method can be used in areas where aerial application is not available or feasible. It is encouraging that this microfibre technique is already registered against one important agricultural pest, the pink bollworm, *Pectinophora gossypiella* (Saunders). This non-disruptive method may well be available to growers for codling moth control in the years to come.

SUMMARY AND CONCLUSIONS

There was full awareness in the 1940's of the potential of integrating chemical and biological control in walnut orchards and some temporary progress was made by minimizing dosages of DDT to achieve a degree of selectivity favoring beneficial species. The biological control component was further strengthened when climatically adapted strains of an introduced parasite of the walnut aphid were successfully established. As a consequence, chemical control of this aphid is necessary in less than 5% of orchards, outbreaks of secondary pests became less frequent in orchards of varieties resistant to codling moth, and chances to control other walnut pests with biological agents improved. Efforts continue to develop adequate biological control of other insect pests of walnuts, with ongoing research and exploration for the parasites of the walnut husk fly, navel orangeworm, and the dusky-veined aphid. Compared to some fruit crops, insecticide and fungicide use in walnut orchards is considerably lower which favors biological control. However, upsets do occur, most often caused by the use of broad-spectrum insecticides against the codling moth, at present necessary in the absence of well developed specific methods of control. The stability of the IPM program on walnuts will increase as broad-spectrum compounds are discontinued and replaced by selective compounds or alternative tactics. Pesticide use on walnuts may be further reduced as new quantitative methods of pest population assessment and spray timing become available. At present the quantitative foundation of walnut pest management is not adequate. Much work remains to be done in the area of monitoring, predictive modeling and the relationship of pest density to nut quality and yield before pest managers can be more certain of the timing, need, and outcome of control actions.

Long term breeding programs on walnuts should take varietal resistance to specific pests into account. Before this can be done, however, more information must be gathered on genetic and ecological resistance mechanisms in walnuts against pest attack. Cultural practices should also be evaluated in the context of walnut pest management seeking to make the orchard environment less favorable for pest organisms, more favorable for their natural enemies, while sound from a horticultural and economic point of view.

Walnut culture in California has been experiencing significant changes over the past 25 years which also bear on pest management. There have been large shifts in walnut acreage from southern and coastal California to the interior valleys, primarily the San Joaquin valley with the heaviest concentration of walnut growing. As a consequence there has been a shift in emphasis in pest management problems to a pest species complex which is more tolerant of the arid climatic conditions in the central valleys of California. Also, parallel to this change in the distribution of walnut acreage in the state there has been a trend to the early, heavy-bearing varieties. Therefore, pests which are better synchronized with or prefer early varieties have become more important.

ACKNOWLEDGMENTS

The authors are grateful for the helpful suggestions offered by the following individuals during the preparation of this manuscript: W. W. Barnett, L. B. Fitch, L. C. Hendricks, A. H. Lange, W. H. Olson, D. E. Ramos, and G. S. Sibbett, all with the Cooperative Extension Service, University of California.

REFERENCES

Allen, M. W. 1952. Root-lesion nematode. *Diam. Wal. News* **34(5)**:8-9.

Anonymous. 1973a. Navel orangeworm, a growing threat to the nut industry. *Western Fruit Grower* **27(6)**:8, 13, 28.

Anonymous. 1973b. Black walnut as a crop. Proc. Black Walnut Symp., Carbondale, Ill., Aug. 14-15, 1973. *USDA Forest Service, Gen. Tech. Rept.* NC-4: 114 pp.

Ark, P. A., and C. E. Scott. 1951. Walnut blight. *Calif. Agric.* **5(3)**:7, 14.

Barnes, M. M. 1978. Unpublished data. Riverside, CA.

Barnes, M. M., and C. S. Davis. 1972. Codling moth control. *Diam. Wal. News* **54(1)**:12, 23.

Barnes, M. M., and H. R. Moffitt. 1978. A five-year study of the effects of the walnut aphid and the European red mite on Persian walnut productivity in coastal orchards. *J. Econ. Entomol.* **71**:71-74.

Barnes, M. M., and J. C. Ortega. 1958. Codling moth in southern California. *Diam. Wal. News* **40(2)**:6, 25.

Barnes, M. M., and J. C. Ortega. 1959. Experiments with protein hydrolysate bait sprays for control of the walnut husk fly. *J. Econ. Entomol.* **52**:279-285.

Barnes, M. M., and H. T. Osborn. 1958. Attractants for the walnut husk fly. *J. Econ. Entomol.* **51**:686-689.

Barnes, M. M., M. J. Garber, and H. R. Moffitt. 1961. Continuing walnut aphid research. *Diam. Wal. News* **43(3)**:19-21.

Barnes, M. M., G. S. Sibbett, and C. S. Davis. 1971. Walnut aphid management: aphid effects on walnut production and quality. *Calif. Agric.* **25(5)**:12-13.

Barnes, M. M., W. C. Batiste, C. S. Davis, and A. S. Deal. 1969. Pest and disease control program for walnuts. *Calif. Agr. Exp. Sta. Ext. Serv.* 19 pp.

Barnes, M. M., C. S. Davis, G. S. Sibbett, and W. W. Barnett. 1978. Integrated pest management in walnut orchards. *Calif. Agric.* **32(2)**:14-15.

Barnes, M. M., *et al.* 1973. Pest and disease control program for walnuts. *Calif. Agr. Exp. Sta. Ext. Serv.* 22 pp.

Barrett, R. E. 1932. An annotated list of the insects and arachnids affecting the various species of walnuts or members of the genus *Juglans* L. *Univ. Calif. Pub. Entomol.* **5**:275-309.

Bartlett, B. R., and J. C. Ortega. 1952. Relation between natural enemies and DDT-induced increases in frosted scale and other pests of walnuts. *J. Econ. Entomol.* **45**:783-785.

Batchelor, L. D. 1929. Walnut culture in California. *Univ. Calif. Agr. Exp. Sta. Bull.* 379. 110 pp. (revised ed.).

Batiste, W. C. 1972. Integrated control of codling moth on pears in California: a practical consideration where moth activity is under surveillance. *Environ. Entomol.* **1**:213-218.

Baumhover, A. H., C. N. Husman, C. C. Skipper, and W. D. New. 1959. Field observations on the effects of releasing sterile screw-worms in Florida. *J. Econ. Entomol.* **52**:1202-1206.

Begg, E. L. 1977. Identification and evaluation of soils for walnuts. pp. 17-21 *In* "Walnut orchard management, Short Course Proceedings." Dept. of Pomology, Univ. of Calif., Davis. 149 pp.

Boyce, A. M. 1932. The walnut husk fly. *Diam. Wal. News* **14(3)**:15-17.

Boyce, A. M. 1934. Bionomics of the walnut husk fly. *Hilgardia* **8(11)**:363-579.

Boyce, A. M. 1935. The codling moth in Persian walnuts. *J. Econ. Entomol.* **28**: 864-873.

Boyce, A. M. 1936. Control of the husk fly. *Diam. Wal. News* **18(4)**:16.

Boyce, A. M. 1946. Summary of studies on control of codling moth on walnuts in southern California. *Diam. Wal. News* **28(3)**:12-23.

Boyce, A. M., and B. R. Bartlett. 1939. Red spiders on walnuts and their control. *Diam. Wal. News* **21(4)**:4-5.

Boyce, A. M., and B. R. Bartlett. 1941a. Lures for the walnut husk fly. *J. Econ. Entomol.* **34**:318.

Boyce, A. M., and B. R. Bartlett. 1941b. Control of codling moth. *Diam. Wal. News* **23(3)**:4-7.

Boyce, A. M., D. T. Prendergast, J. F. Kagy, and J. W. Hansen. 1939. Dinitro-o-cyclohexylphenol in the control of mites on citrus and Persian walnuts. *J. Econ. Entomol.* **32**:450-467.

Bruce, D. L., and M. M. Barnes. 1973. Navel orangeworm still defies control. *Diam. Wal. News* **55(1)**:15, 30.

Buckingham, G. R. 1975. The parasites of walnut husk flies (Diptera:Tephritidae: Rhagoletis) including comparative studies on the biology of *Biosteris juglandis* Mues. (Hymenoptera: Braconidae) and on the male tergal glands of the Braconidae (Hymenoptera). Ph.D. Thesis, University of California, Berkeley.

Butt, B. A. 1975. Bibliography of the codling moth. *USDA - ARS* W-31:221 pp.

Butt, B. A., L. D. White, H. R. Moffitt, C. O. Hathaway, and L. G. Schoenleber. 1973. Integration of sanitation, insecticides, and sterile moth releases for suppression of populations of codling moths in the Wenas valley of Washington. *Environ. Entomol.* **2**:208-212.

Caltagirone, L. E. 1966. A new *Pentalitomastix* from Mexico. *Pan-Pacific Entomol.* **42(2)**:145-151.

Campbell, R. E. 1923. Notes on nicotine dust progress. *J. Econ. Entomol.* **16**: 497-505.

Chace, E. M. 1935. The ethylene process and its place in walnut harvesting. *Diam. Wal. News* **17(2)**:19.

Chandler, W. H. 1957. Deciduous Orchards. Lee and Febiger Co., Philadelphia, 492 pp. (3rd ed.).

Collins, D. L., and W. Machado. 1937. Effects of light traps on a codling moth infestation: a consideration of four years' data. *J. Econ. Entomol.* **30**: 422-427.

Croft, B. A. 1975a. Integrated control of orchard pests in the USA. *C. R. 5e Symp. Lutte Integree en vergers. OILB/SROP* 1975: 109-124.

Croft, B. A. 1975b. Tree fruit pest management. *In* "Introduction to insect pest management." (R. L. Metcalf and W. H. Luckmann, eds.). Wiley Interscience, 587 pp.

Croft, B. A. 1975c. Integrated control of apple mites. *Mich. State Univ. Ext. Bull.* E-825: 12 pp.

Croft, B. A., S. M. Welch, and M. J. Dover. 1976. Dispersion statistics and sample size estimates for populations of the mite species *Panonychus ulmi* and *Amblyseius fallacis* on apples. *Environ. Entomol.* **5**:227-234.

Culver, D. J. 1978a. Insect management makes more profit (Part I). *Diam. Wal. News* **60(1)**:5-7.

Culver, D. J. 1978b. Insect management makes more profit (Part II). *Diam. Wal. News* **60(2)**:5-8.

Davis, C. S., R. Hom, R. van den Bosch, M. M. Barnes, G. S. Sibbett, R. B. Jeter, and F. M. Charles. 1972. Impact on *Trioxys pallidus. Diam. Wal. News* **54(1)**:8, 23.

Day, B. E., L. S. Jordan, O. D. Mann, and R. C. Russell. 1964. Weed control in walnuts. *Diam. Wal. News* **46(6)**:6, 7, 16.

Elmore, C. L., and A. H. Lange. 1977. Vegetation management in walnuts. pp. 91-94 *In* "Walnut orchard management, Short Course Proceedings." Dept. of Pomology, Univ. of Calif., Davis. 149 pp.

Etzel, L. 1978. Personal communication to Helmut Riedl.

Falcon, L. A., C. Pickel, and J. White. 1975. Computerizing codling moth. *Western Fruit Grower* **96(1)**:8-14.

Fawcett, H. S., and L. D. Batchelor. 1920. An attempt to control walnut blight. *Calif. Dept. Agr. Monthly Bull.* **9**:5-6.

Fereres, E. 1977. Scheduling irrigations in walnuts based on evaporation measurements. *In* "Walnut orchard management, Short Course Proceedings." Dept. of Pomology, Univ. of Calif., Davis. 149 pp.

Finch, C. U., and W. C. Sharp. 1976. Cover crops in California orchards and vineyards. USDA, Soil Conservation Service, 25 pp.

Fischer, B. B., A. H. Lange, J. McCaskill, and B. Crampton. 1976. Growers' weed identification handbook. *Univ. Calif. Agr. Exp. Sta. Ext. Serv. Pub.* 4030. (rev. ed.).

Fitz, J., and J. W. Fitz. 1872. The southern apple and peach culturist. J. W. Randolph and English, Richmond, Virginia.

Foster, S. W. 1912. On the nut-feeding habits of the codling moth. *USDA Bur. Ent. Bull.* **80(5)**:67-70.

Frick, K. E. 1952. Determining emergence of the cherry fruit fly with ammonium carbonate bait traps. *J. Econ. Entomol.* **45**:262-263.

Gerhardt, P. P., D. L. Lindgren, and W. B. Sinclair. 1951. Methyl bromide fumigation of walnuts to control two lepidopterous pests, and determination of bromine residue in walnut meats. *J. Econ. Entomol.* **44**:384-389.

Giese, R. L., R. M. Peart, and R. M. Huber. 1975. Pest management – a pilot project exemplifies new ways of dealing with important agricultural pests. *Sci.* **187**:1045-1052.

Hagen, K. S. 1978. Personal communication to Helmut Riedl.

Haynes, D. L., R. K. Brandenburg, and P. D. Fisher. 1973. Environmental monitoring network for pest management systems. *Environ. Entomol.* **2**:889-899.

Hendricks, L. C., and H. I. Forde. 1977a. Selection of walnut varieties. *Diam. Wal. News* **59(6)**:20-23.

Hendricks, L. C., and H. I. Forde. 1977b. Selection of walnut varieties. Pp. 39-43 *In* "Walnut orchard management, Short Course Proceedings." Dept. of Pomology, Univ. of Calif., Davis. pp. 149.

Herms, W. B. 1932. Deterrent effect of artificial light on the codling moth. *Hilgardia* 7(7):263-280.

Hodgson, R. W. 1923. The nematode a serious pest of the walnut. *Diam. Wal. News* 5(4):6.

Hodson, A. C. 1948. Further studies of lures attractive to the apple maggot. *J. Econ. Entomol.* **41**:61-66.

Hoyt, S. C. 1969. Integrated chemical control of insects and biological control of mites on apple in Washington. *J. Econ. Entomol.* **62**:74-86.

Huber, J., and E. Dickler. 1977. Codling moth granulosis virus: Its efficacy in the field in comparison with organophosphorus insecticides. *J. Econ. Entomol.* **70**:557-561.

Husseiny, M. M., and H. F. Madsen. 1964. Sterilization of the navel orangeworm, *Paramyelois transitella* (Walker), by gamma radiation (Lepidoptera: Phycitidae). *Hilgardia* **35(3)**:113-137.

Jones, A. L. 1976. Apple scab predictive and monitoring systems. Proc. US-USSR Symp.: The Integrated Control of the Arthropod, Disease and Weed Pests of Cotton, Grain Sorghum and Deciduous Fruit, Sept. 28-Oct. 1, 1975, Lubbock, Texas: 78-84.

Joos, J. L., C. S. Davis, and F. M. Charles. 1974. Walnut husk fly trapping procedures. *Univ. Calif. Agr. Ext. Serv.*, OSA No. 176.

Lange, A. H. 1968. Weeds in California fruit crops: a summary of problems and herbicide possibilities. *Calif. Agric.* **22(2)**:8-10.

LaRue, J. H., A. O. Paulus, W. D. Wilbur, H. J. O'Reilly, and E. F. Darley. 1962. *Armillaria* root fungus. *Diam. Wal. News* **44(3)**:8, 17.

Lownsbery, B. F. 1977. Nematodes parasitizing walnut roots and their control. Pp. 73-75 *In* "Walnut orchard management, Short Course Proceedings." Dept. of Pomology, Univ. of Calif., Davis. 149 pp.

Lownsbery, B. F., and S. A. Sher. 1958. Root-lesion nematode on walnut. *Calif. Agric.* **12(5)**:7, 12.

Lownsbery, B. F., W. H. Hart, and G. C. Martin. 1969. Root-lesion nematodes. *Diam. Wal. News* **51(1)**:20-22.

MacLellan, C. R. 1976. Suppression of codling moth (Lepidoptera: Olethreutidae) by sex pheromone trapping of males. *Can. Entomol.* **108**:1037-1040.

Madsen, H. F., and J. M. Vakenti. 1972. Codling moths: female-baited and synthetic pheromone traps as population indicators. *Environ. Entomol.* **1**:554-557.

Madsen, H. F., and J. M. Vakenti. 1973. Codling moth: use of Codlemone®-baited traps and visual detection of entries to determine need of sprays. *Environ. Entomol.* 2:677-679.

Madsen, H. F., and T. T. Y. Wong. 1962. Tracking the navel orangeworm. *Diam. Wal. News* **44(3)**:8, 14.

Madsen, H. F., L. A. Falcon, and T. T. Y. Wong. 1964. Control of walnut aphid and codling moth on walnuts in northern California. *J. Econ. Entomol.* **57**:950-952.

Madsen, H. F., J. M. Vakenti, and F. E. Petero. 1976. Codling Moth: suppression by male removal with sex pheromone traps in an isolated apple orchard. *J. Econ. Entomol.* **69**:597-599.

Martin, G. C., and H. I. Forde. 1975. Blackline. *Diam. Wal. News* **57(5)**:13-14.

McEwen, F. L., E. H. Glass, A. C. Davis, and C. M. Splittstoesser. 1960. Field tests with *Bacillus thuringiensis* Berliner for control of four lepidopterous pests. *J. Insect Pathol.* **2**:152-164.

Meyer, J. L., F. K. Aljibury, and J. A. Beutel. 1975. Postharvest irrigation of orchards. *Univ. Calif. Coop. Ext. Leaf.* 2767: 2 pp.

Michelbacher, A. E. 1945. The importance of ecology in insect control. *J. Econ. Entomol.* **38**:129-130.

Michelbacher, A. E. 1953. New materials for aphid control. *Diam. Wal. News* **35(3)**:8.

Michelbacher, A. E. 1954. Spider mites in northern California. *Diam. Wal. News* **36(4)**:20.

Michelbacher, A. E. 1955. Frosted scale on walnuts in northern California. *Pan-Pacific Entomol.* **31(3)**:139-148.

Michelbacher, A. E. 1956. Calico scale on walnuts. *Diam. Wal. News* **38(6)**:15.

Michelbacher, A. E. 1957. Aphid control essential. *Diam. Wal. News* **39(2)**:12-13.

Michelbacher, A. E., and O. G. Bacon. 1952a. Walnut insect and spider mite control in northern California. *J. Econ. Entomol.* **45**:1020-1027.

Michelbacher, A. E., and O. G. Bacon. 1952b. Spider mites on walnuts. *Calif. Agric.* **6(6)**:4, 15.

Michelbacher, A. E., and C. S. Davis. 1961a. Navel orangeworm on walnuts in northern California. *Calif. Agric.* **15(9)**:12, 13.

Michelbacher, A. E., and C. S. Davis. 1961b. The navel orangeworm in northern California. *J. Econ. Entomol.* **54**:559-562.

Michelbacher, A. E., and W. W. Middlekauff. 1948. Control of codling moth on Payne walnuts in northern California. *Diam. Wal. News* **31(2)**:6-8.

Michelbacher, A. E., and W. W. Middlekauff. 1949. Codling moth investigations on its Payne variety of English walnut in northern California. *J. Econ. Entomol.* **42**:736-746.

Michelbacher, A. E., and J. C. Ortega. 1958. A technical study of insects and related pests attacking walnuts. *Calif. Agr. Exp. Sta. Bull.* 764. 86pp.

Michelbacher, A. E., and R. L. Sisson. 1956. The walnut husk fly – pest new in northern California found in Sonoma Valley near Santa Rosa. *Calif. Agric.* **10(5)**:13.

Michelbacher, A. E., and C. Swanson. 1945. Factors influencing the control of walnut aphid. *J. Econ. Entomol.* **38**:127-128.

Michelbacher, A. E., and C. Swanson. 1946. Control of codling moth on Payne walnuts in the San Joaquin Valley. *Diam. Wal. News* **28(3)**:6-7.

Michelbacher, A. E., and J. E. Swift. 1954. Parasites of the frosted scale. *Diam. Wal. News* **36(3)**:16-17.

Michelbacher, A. E., S. Hitchcock, and A. H. Retan. 1956. Filbertworm injury to walnuts. *Calif. Agric.* **10(1)**:11, 12.

Michelbacher, A. E., W. W. Middlekauff, and C. Hansen. 1951. Occurrence of a fungus disease in overwintering stages of the codling moth. *J. Econ. Entomol.* **43**:955.

Michelbacher, A. E., W. W. Middlekauff, and C. Swanson. 1947. Insect investigations on Payne walnuts in San Joaquin County. *Diam. Wal. News* **29(3)**: 8-11.

Michelbacher, A. E., W. W. Middlekauff, and E. Wegenek. 1950. Timing of spray treatments for codling moths. *Calif. Agric.* **4(3)**:8-10.

Michelbacher, A. E., A. H. Retan, and S. Hitchcock. 1957. Filbertworm injury to walnuts. *Calif. Agric.* **11(9)**:7.

Michelbacher, A. E., C. Swanson, and G. F. MacLeod. 1944. Control of codling moth on walnuts in San Joaquin County. *Diam. Wal. News* **26(3)**:4-6.

Michelbacher, A. E., W. W. Middlekauff, D. Davis, and C. Cassill. 1950. Control of codling moth in northern California. *Diam. Wal. News* **32(2)**:4-5.

Middlekauff, W. W., and A. E. Michelbacher. 1951. Spider mite infestation on walnuts. *Diam. Wal. News* **33(1)**:6-9.

Mircetich, J. S. M., W. J. Moller, and D. E. Ramos. 1977a. Root and crown rot diseases of walnut trees. pp. 58-68 *In* "Walnut orchard management, Short Course Proceedings." Dept. of Pomology, Univ. of Calif., Davis. 149 pp.

Mircetich, J. S. M., W. J. Moller, and R. R. Sanborn. 1977b. Blackline disease of walnut trees. pp. 69-72 *In* "Walnut orchard management, Short Course Proceedings." Dept. of Pomology, Univ. of Calif., Davis. 149 pp.

Moller, W. J., and M. N. Schroth. 1976. Biological control of crown gall. *Calif. Agric.* **30(8)**:8-9.

New, P. B., and A. Kerr. 1972. Biological control of crown gall – field measurements and glasshouse experiments. *J. Appl. Bact.* **35**:279-287.

Nickel, J. L., and T. T. Y. Wong. 1966. Control of the walnut husk fly, *Rhagoletis completa* Cresson, with systemic insecticides. *J. Econ. Entomol.* **59**: 1079-1082.

Nixon, P. L., and J. E. McPherson. 1977. An annotated list of phytophagous insects collected on immature black walnut trees in southern Illinois. *Great Lakes Entomol.* **10(4)**:211-222.

Olson, W. H. 1974. Dusky-veined walnut aphid studies. *Calif. Agric.* **28(7)**:18-19.

Olson, W. H., W. J. Moller, L. B. Fitch, and R. B. Jeter. 1976. Walnut blight control. *Calif. Agric.* **30(5)**:10-13.

Olson, W. H., L. C. Hendricks, G. S. Sibbett, C. S. Davis, and D. E. Ramos. 1975. Navel orangeworm control through early harvest. *Calif. Agric.* **29(9)**:3.

Ortega, J. C. 1948. DDT controls codling moth. *Diam. Wal. News* **30(2)**:6-8.

Ortega, J. C. 1950a. Codling moth control in southern Calif. *Diam. Wal. News* **32(3)**:4-5.

Ortega, J. C. 1950b. The navel orangeworm on walnuts in southern California. *Diam. Wal. News* **32(5)**:6-7.

Ortega, J. C. 1956a. DDT plus systemics controls codling moths, aphids and mites. *Diam. Wal. News* **38(2)**:6-8.

Ortega, J. C. 1956b. Walnut husk fly control. *Diam. Wal. News* **38(4)**:14.

Osgood, J. W. 1977. Soil management and cultural systems. pp. 89-90 *In* "Walnut orchard management, Short Course Proceedings." Dept. of Pomology, Univ. of Calif., Davis. 149 pp.

Phillips, P. A., and M. M. Barnes. 1975. Host race formation among sympatric apple, walnut and plum populations of the codling moth, *Laspeyresia pomonella. Ann. Entomol. Soc. Am.* **68**:1053-1060.

Proverbs, M. D., D. M. Logan, and J. R. Newton. 1975. A study to suppress codling moth (Lepidoptera: Olethreutidae) with sex pheromone traps. *Can. Entomol.* **107**:1265-1269.

Proverbs, M. D., J. R. Newton, and D. M. Logan. 1966. Orchard assessment of the sterile male technique for control of the codling moth, *Carpocapsa pomonella* (L.) (Lepidoptera: Olethreutidae). *Can. Entomol.* **98**:90-95.

Proverbs, M. D., J. R. Newton, and D. M. Logan. 1977. Codling moth control by the sterility method in twenty-one British Columbia orchards. *J. Econ. Entomol.* **70**:667-671.

Quayle, H. J. 1926. The codling moth in walnuts. *Calif. Agr. Exp. Sta. Bull.* 402. 33 pp.

Rackham, R. L., and H. J. O'Reilly. 1967. Crown rot of deciduous fruit and nut trees. *Univ. Calif. Agric. Exp. Sta. Leaf.* 195.

Rice, R. E. 1978. Personal communication to Helmut Riedl.

Rice, R. E., L. L. Sadler, M. L. Hoffman, and R. A. Jones. 1976. Egg traps for the navel orangeworm, *Paramyelois transitella* (Walker). *Environ. Entomol.* **5**:697-700.

Riedl, H. 1979. Monitoring and forecasting methods for codling moth management in North America. *EPPO Bull.* (in press).

Riedl, H., and B. A. Croft. 1974. A study of pheromone trap catches in relation to codling moth damage. *Can. Entomol.* **106**:525-537.

Riedl, H., and B. A. Croft. 1978a. Management of the codling moth in Michigan. *Res. Rept. Mich. State Univ. Agr. Exp. Sta. (Farm Science)* 337: 18 pp.

Riedl, H., and B. A. Croft. 1978b. The effect of photoperiod and effective temperatures on the seasonal phenology of the codling moth. *Can. Entomol.* **110**:455-470.

Riedl, H., B. A. Croft, and A. J. Howitt. 1976. Forecasting codling moth phenology based on pheromone trap catches and physiological time models. *Can. Entomol.* **108**:449-460.

Rizzi, A. D. 1977. History, geographic distribution, and climatic adaptation of walnuts. pp. 4-9 *In* "Walnut orchard management, Short Course Proceedings." Dept. of Pomology, Univ. of Calif., Davis. 149 pp.

Rock, R. C., and A. D. Rizzi. 1955. The where and when of California fruit and nut crops. *Univ. of Calif. Agr. Exp. Sta. Manual 20.*

Ross, N., M. N. Schroth, R. Sanborn, H. J. O'Reilly, and J. P. Thompson. 1970. Reducing loss from crown gall disease. *Univ. Calif. Agr. Exp. Sta. Bull.* 845: 10 pp.

Rudolph, B. A. 1933. Bacteriosis of the English walnut and its control. *Calif. Agr. Exp. Sta. Bull.* 564:1-88.

Schaad, N. W., and E. E. Wilson. 1971. Bacterial phloem canker of Persian walnut. *Calif. Agric.* **25(4)**:4-7.

Schlinger, E. I., K. S. Hagen, and R. van den Bosch. 1960. Imported French parasite of walnut aphid established in California. *Calif. Agric.* **14(11)**:3-4.

Schroth, M. N. 1978. Personal communication to Helmut Riedl.

Schroth, M. N., and E. Mulrean. 1977. Understanding the disease cycle of walnut blight. *Diam. Wal. News* **59(5)**:12, 13, 48.

Schroth, M. N., E. Mulrean, W. J. Moller, and B. Teviotdale. 1977. Bacterial diseases of walnuts. pp. 135-139 *In* "Walnut orchard management, Short Course Proceedings." Dept. of Pomology, Univ. of Calif., Davis. 149 pp.

Serr, E. F. 1959. Blackline in walnut. *Calif. Agric.* **13(3)**:8, 9, 14, 15.

Serr, E. F., and J. P. Fairbank. 1944. Mechanical harvesting of walnuts is making progress. *Diam. Wal. News* **26(3)**:8-10.

Serr, E. F., and H. I. Forde. 1968. Ten new walnut varieties released. *Calif. Agric.* **22(4)**:8-10.

Sibbett, G. S., and C. S. Davis. 1978. Insect management. *Diam. Wal. News* **60(2)**:9-10.

Sibbett, G. S., C. S. Davis, and M. M. Barnes. 1971. Walnut aphid and the sunburn problem. *Calif. Agric.* **25(5)**:13-14.

Sibbett, G. S., F. J. Perry, L. C. Hendricks, W. R. Schreader, and R. H. Tyler. 1972. Ethrel, a harvest aid for walnuts. *Diam. Wal. News* **54(4)**:8-11.

Sibbett, G. S., L. C. Hendricks, G. Carnill, W. H. Olson, R. Jeter, D. E. Ramos, G. C. Martin, and C. S. Davis. 1974. Walnut quality and value maximized by harvest management. *Calif. Agric.* **28(7)**:15-17.

Sluss, R. R. 1967. Population dynamics of the walnut aphid, *Chromaphis juglandicola* (Kalt.) in Northern California. *Ecology* **48(1)**:41-57.

Smith, R. E. 1921. The preparation of nicotine dust as an insecticide. *Calif. Agric. Exp. Sta. Bull.* 336:261-274.

Smith, R. E. 1953. 50th Anniversary of Plant Pathology Dept. *Diam. Wal. News* **35(4)**:10.

Teviotdale, B. L., and W. J. Moller. 1977. Disease control guidelines for walnuts. *Mimeo. Div. Agric. Sci., Univ. of Calif.* 7 pp.

Thomson, S. V., M. N. Schroth, W. J. Moller, W. O. Reid, J. A. Beutel, and C. S. Davis. 1977. Pesticide applications can be reduced by forecasting the occurrence of fireblight bacteria. *Calif. Agric.* **31(10)**:12-14.

Tummala, L. T., and D. L. Haynes. 1977. On-line pest management systems. *Environ. Entomol.* **6**:339-349.

Van den Bosch, R. 1978. Personal communication to Helmut Riedl.

Van den Bosch, R., E. I. Schlinger, and K. S. Hagen. 1962. Initial field observations in California on *Trioxys pallidus* (Haliday), a recently introduced parasite of the walnut aphid. *J. Econ. Entomol.* **55**:857-862.

Van den Bosch, R., B. D. Frazer, C. S. Davis, P. S. Messenger, and R. Horn. 1970. An effective new walnut aphid parasite from Iran. *Calif. Agric.* **24(11)**: 8-10.

Yothers, M. A. 1930. Summary of results obtained with trap baits in capturing the codling moth in 1927. *J. Econ. Entomol.* **20**:576-587.

STATUS OF PEST MANAGEMENT PROGRAMS FOR THE PECAN WEEVIL

David J. Boethel
Louisiana Agricultural Experiment Station
Pecan Research and Extension Station
Shreveport, Louisiana

R. D. Eikenbary
Department of Entomology
Oklahoma State University
Stillwater, Oklahoma

INTRODUCTION

The pecan, *Carya illinoensis* Koch, is the most important native horticultural crop in the United States; however, the pecan industry is new compared to other orchard crops with commercial pecan production developing largely during the 20th century (Brison 1974). Even today native (seedling) pecans comprise a substantial portion of the crop that enters commerce. Because the trees that bear these nuts are generally found growing randomly along river valleys, acreage figures are difficult to attain. Therefore, the magnitude of the commodity can best be expressed as pounds of inshell pecans produced.

The average annual production of pecans for the years 1971-77 was 203.8 million pounds at an average annual value of $92.8 million. Approximately 57% of the nuts marketed during this period were improved variety (grafted) pecans with the remaining production coming from native pecans (Anon. 1977). In comparison, the average production from 1919-30 was less than 50 million pounds a year (Brison 1974).

The major producing states in the native range of pecans are Arkansas, Louisiana, Mississippi, Oklahoma, and Texas. These states also producc improved pecans along with Alabama, Florida, South Carolina, and Georgia with the latter being the leading state in annual total production. Relatively recent pecan producers are the states of New Mexico and Arizona which along with the El Paso Valley of Texas make up the "western irrigated region." Plantings have been established in California but total acreage was below 1000 acres in 1976 (Johnson 1978).

As the case with other orchard crops, the pecan has rather exacting requirements for growth and production. For a detailed discussion of the soil and climatic requirements of pecans, cultural requirements, and related subjects, the readers are referred to Brison's 1974 book, *Pecan Culture.*

Successful pest control is no less important to growth and production. The pecan is under attack from various insects, mites, and diseases from bud-break until shuck-split, a period of *ca.* seven months. Thus it is understandable that the escalating costs of production and concern for the environment have focused attention on the development of pest management systems that increase the efficiency of chemical control (predominant contemporary management tactic) and eventually reduce the reliance on pesticides as the major management tool.

Among the insect pests receiving special attention is the pecan weevil, *Curculio caryae* (Horn), generally considered to be the most important arthropod pest of pecans (Fig. 1). The pecan weevil certainly fits the definition of a key pest proposed by Smith and Reynolds (1966) because it is a serious, perennially occurring, persistent pest that dominates late season control programs. In the absence of controls, infestations can cause severe damage leading to substantial economic loss for producers. The purpose of this chapter is to review the status of pest management research with respect to this insect.

THE PECAN WEEVIL

Damage

The pecan weevil causes two types of direct damage to pecan nuts. Prior to shell hardening, feeding punctures by adults will cause immature pecans to fall from the trees. This damage is occasionally overlooked or attributed to other insects also known to cause premature nut drop.

The second type of damage occurs later in the season. After shell hardening, females readily oviposit in the pecans, and the resulting larvae completely destroy the kernels by their feeding. This damage is exhibited at harvest either by wormy pecans or nuts that contain larval emergence holes.

As early as 1904, pecan growers in Georgia, Mississippi, and Texas made inquiries about holes in their pecans with some reporting 75% crop loss due to this type injury (Chittenden 1908). Estimated damage from the pecan weevil has ranged from 35 to 50 million pounds annually (Anon. 1970). Barry (1974) reported that the total cost of the pecan weevil to Georgia's pecan growers was approximately \$7.4 million annually, \$4.9 million in damage and \$2.5 million in control costs.

Pecan buyers and processors also are affected by the pecan weevil. Since all infested pecans are not detected at harvest, these are gathered along with sound pecans. Thus the processor incurs additional costs in removing weevil infested

Figure 1. Female pecan weevil on nut cluster.

nuts. To offset these losses, processors and buyers usually pay lower prices for pecans from areas with histories of high pecan weevil infestations or ultimately stop buying pecans from these areas.

Distribution

The pecan weevil is indigenous to North America and according to Gibson (1969), apparently limits its host range to the genus *Carya.* Until recently, the weevil's distribution was considered to be "west from New York to Iowa and south to Oklahoma, Texas and Georgia" (Gibson 1969); however, a more detailed survey for the pest has revealed that it does not occur in localized areas in some of the pecan producing states (Harris 1979).

Since the pecan is a recent introduction into the "western irrigated region," the pecan weevil is not found in this area and quarantine regulations exist to govern shipment of pecans from infested areas into California, Arizona, New Mexico, and far West Texas. These regulations are of recent origin also, with several researchers investigating suitable treatments for post-harvest control (Harris 1973; Payne and Wells 1974; Kirkpatrick *et. al.* 1976; Leesch and Gillenwater 1976).

An isolated infestation of pecan weevil was discovered in Tularosa, New Mexico in 1970 (Anon. 1970). "Eradication?" procedures were initiated emphasizing extensive chemical control and destruction of all pecans from the invested trees for several years. At the writing of this report, no further infestations have been reported in the area.

Harris (1979) reported that pecan weevil infestations have not been confirmed in Mexico. The pest is presently considered absent from that country.

Life History and Habits

Duration of the life cycle. For many years, the duration of the pecan weevil's life cycle was questionable. Early researchers reported a 1 or 2 year cycle (Leiby 1925), a 1 to 3 year cycle (Hinrichs and Thomson 1955), a 2 year cycle (Dupree and Bissell 1965), a 2 or 3 year cycle (Harned 1929; Moznette *et al.* 1931), a 3 year cycle (Swingle 1934; Kumpe and Isley 1936), and a 4 year life cycle (Price 1939). Much of the controversy associated with these early reports was clarified by Harp (1970). He observed a cohort of 1200 individuals and found that a majority of the pecan weevil population completed their life cycle in two years with a small fraction of the population requiring three years. The "typical" life history of the pecan weevil is illustrated in Fig. 2. In the sections that immediately follow, the various facets of the life history will be discussed as they relate to management of the pest.

Adult emergence. Early workers on the biology of the pecan weevil established that times of adult emergence from the soil varied from year to year, but usually occurred in late summer or early fall. Moznette *et al.* (1931) were the first researchers who attempted to associate a climatic phenomenon (precipitation) with the onset of weevil emergence. Price (1939), Hinrichs (1948), Nickels (1950), and Dupree and Beckham (1953) observed that rainfall appeared to stimulate emergence. Dupree and Bissell (1965) reported that over a 10-year period peak emergence appeared to occur in late August during years of normal rainfall. The years in which variations from this pattern occurred were associated with periods of drought preceding emergence.

Although the early literature linked weevil emergence with rainfall, much of the information was speculative rather than based on quantitative data. Hinrichs and Thomson (1955) were the first to gather empirical data concerning amounts of rainfall and corresponding weevil emergence. Over a span of four years in Oklahoma, they found that emergence occurred earlier in the season following rainfall in late July and early August. Data by Raney *et al.* (1970), also from Oklahoma, confirmed that of Hinrichs and Thomson, and they stated "weevil emergence increased 3-4 days after a 1-2 inch rainfall." In Georgia, Tedders (1974) reported that emergence commenced two days following a two inch rainfall. Prior to that time, the soil moisture had been low (*ca.* 18.5%). Under those conditions, the soil became hard and may have delayed weevil emergence.

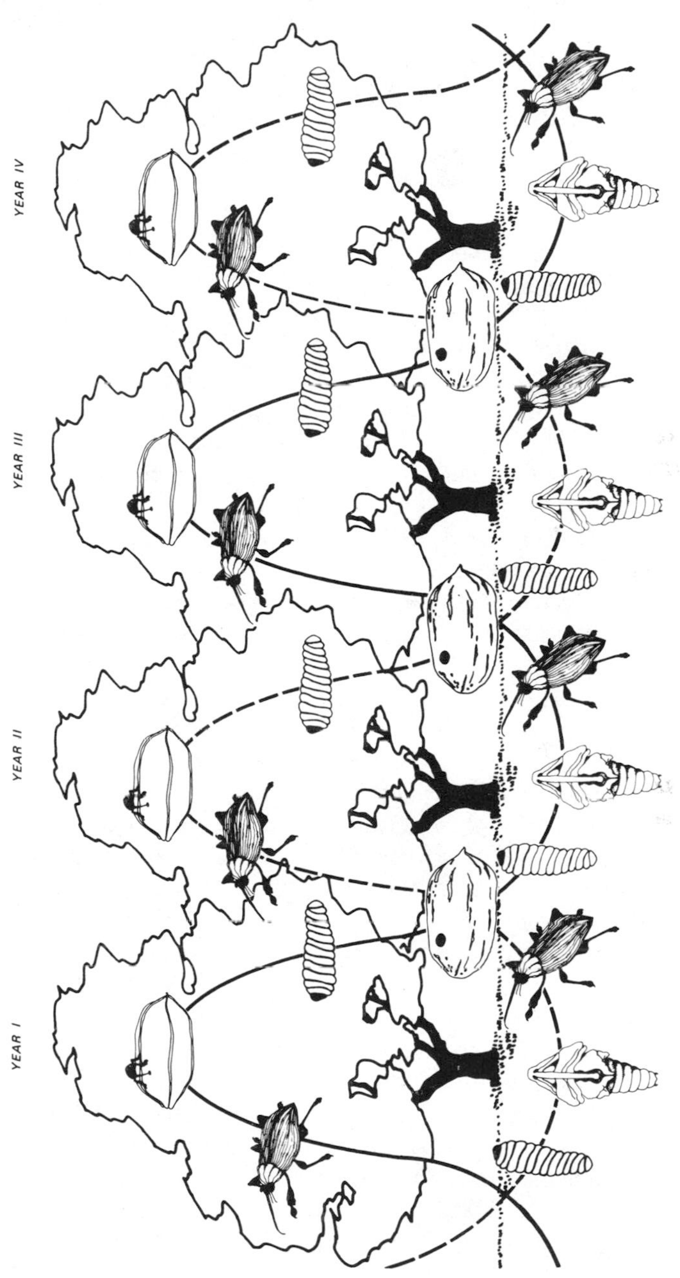

Figure 2. Typical pecan weevil life cycle. Taken from: McWhorter *et al.* (1976).

Harp (1970) studied rainfall and emergence at two locations in Texas. He reported that rainfall may influence emergence but was not convinced that emergence was totally moisture dependent. Since the pecan weevil had been an adult approximately a year before leaving the soil, it was unusual that emergence did not occur after rains earlier in the year.

Boethel (1974) also reported conflicting data concerning rainfall and emergence in two years of study in Oklahoma. In 1972, peak emergence occurred during the last week of August two to three days following the first substantial rainfall of the month (Fig. 3). Prior to this time, only incidental emergence had occurred and the orchard soil had become extremely compact making the taking of soil samples for moisture determination difficult. However, in 1973, peak emergence occurred during the first two weeks in September although rainfall had occurred early in August (Fig. 4). In fact, soil moisture levels on August 9 in 1973 were equal to those encountered in 1972 when peak emergence took place. Harp (1970) hypothesized that rainfall might suppress weevil emergence, and Boethel (1974) observed that on several days when rainfall occurred emergence was depressed (Fig. 4).

Neel *et al.* (1975) found that adult emergence was delayed into late September and October in Arkansas when drought conditions existed in August. Harris (1978) obtained similar results in Texas where peak emergence occurred in October following lack of precipitation in August and September.

Studies conducted at four locations for four years in Louisiana have demonstrated that peak emergence usually takes place during the last week in August and the first week of September (Boethel 1978). These results confirm the findings of Dupree and Bissell (1965). Thus studies indicating late August and early September as the key emergence period give substance to Harp's (1970) suggestion that the pecan weevil had evolved in perfect synchrony with the fruit maturity of its host and emerged each year accordingly. Therefore, it appears rainfall may be linked to emergence only as it affects soil compactibility (permeability) in late season.

Movement into the tree and dispersal. Pecan weevils have definite habits in reaching the trees. Raney and Eikenbary (1968) studied the flight habits of the pest and found that females tended to fly higher than males. Male weevils were observed to be clumsy flyers and crawled to the trees more often than females. Subsequent mark and recapture studies indicated that weevils dispersed from the ground to the top of 35-ft trees in 24 hours and that a tendency existed to move higher in trees that contained fewer nuts (Raney *et al.* 1969). Other research on intratree dispersal demonstrated that weevils moved considerably throughout the canopy regardless of the height or sector of the tree in which released (Eikenbary and Raney 1973). These observations were further confirmed in studies that demonstrated that no intratree differences in nut infestations occurred (Boethel *et al.* 1974).

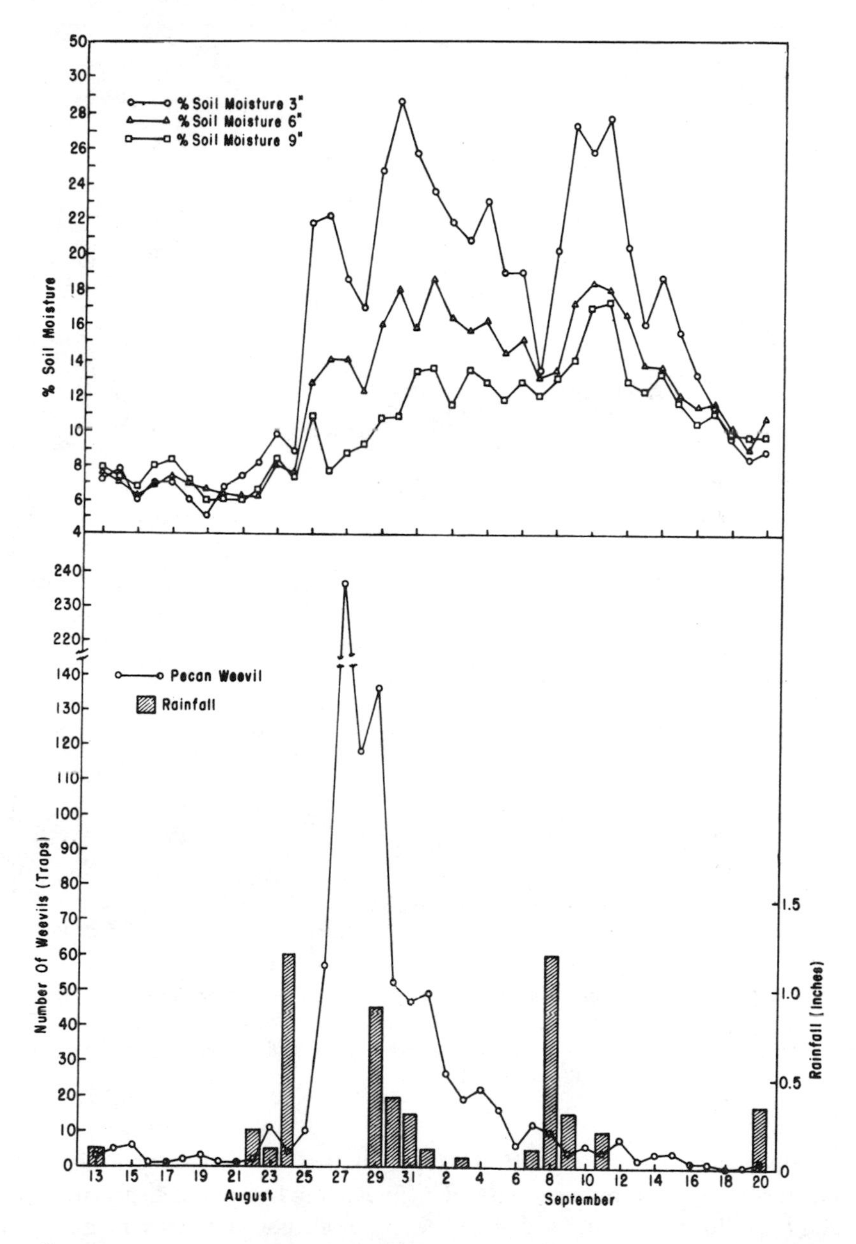

Figure 3. Emergence of pecan weevils at Stillwater, Oklahoma, 1972. The number of weevils shown represents the total collected from 120 traps beneath 10 test trees. The % soil moisture represents the daily mean at each sample depth. Modified from: Boethel (1974).

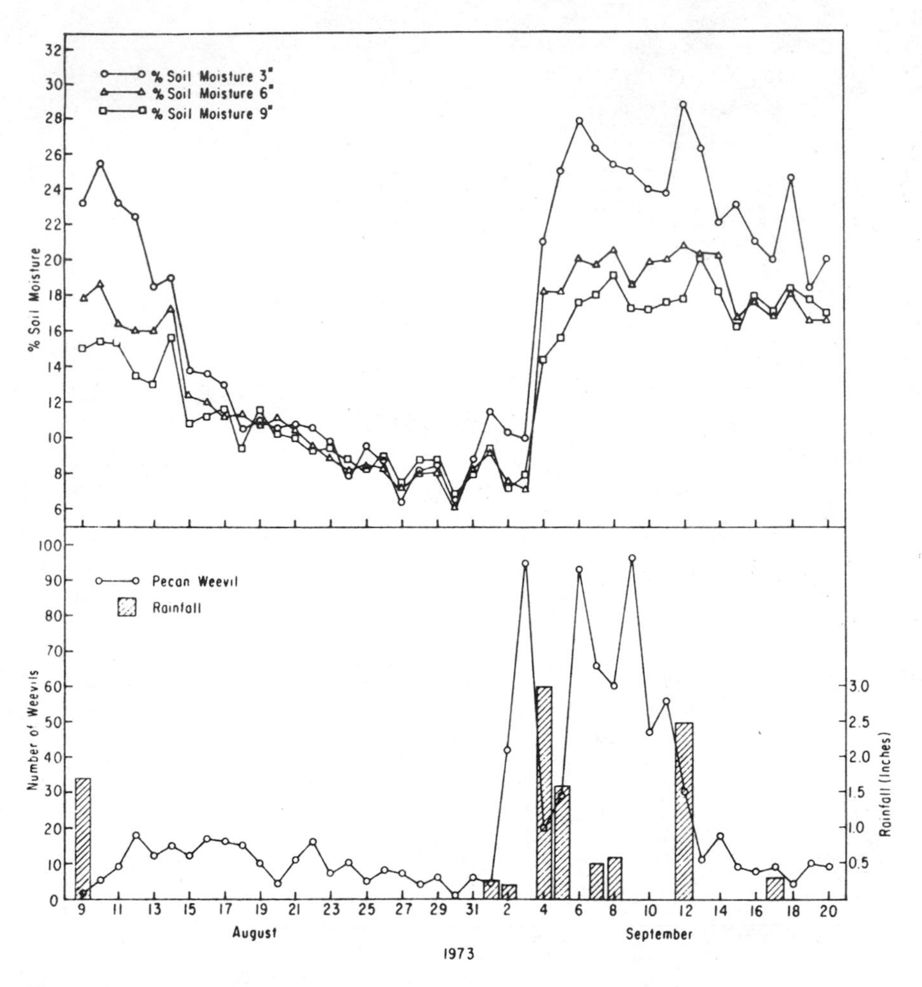

Figure 4. Emergence of pecan weevils at Stillwater, Oklahoma, 1973. The number of weevils shown represents the total collected from 120 traps beneath 10 test trees. The % soil moisture represents the daily mean at each sample depth. Modified from: Boethel (1974).

Eikenbary and Raney (1973) reported that marked weevils were found one-quarter mile from the release point indicating horizontal dispersal. Boethel *et al.* (1976b) also observed intertree dispersal; however, movement between trees occurred late in the season after peak emergence periods.

Ovipositional habits. The relationship of nut naturity and pecan weevil oviposition is well documented in the literature. Moznette *et al.* (1931) were the first to report that oviposition did not begin until the shells were nearly hardened and contained well developed kernels. In recent years, several experiments have confirmed this report (Van Cleave and Harp 1971; Criswell *et al.* 1975; Harris 1976a).

Criswell *et al.* (1975) studied ovipositional activities on three pecan varieties which differed in maturity dates. They found that weevils oviposited on the early maturing variety ('Hardy Giant') earlier than on the late maturing varieties (Fig. 5). Harris (1976b) studied native trees with different phenologies and also found that oviposition occurred earlier on trees with earlier maturing nuts. The importance of this phenomenon in developing a management program for the weevil will be treated in more detail later.

Van Cleave and Harp (1971) stated, "Since the pecan weevil has been an adult for approximately one year before it emerges from the soil, it is difficult to define its preoviposition period." In their study, the preoviposition period was considered to be the time between emergence from the soil and oviposition and was found to be a minimum of 5 days. Criswell *et al.* (1975) observed oviposition as early as 2 days after emergence but the average preovipositional period was 6.5 days. They also found that peak egg production was reached 10-12 days after emergence (Fig. 6).

Swingle (1935) reported that female weevils required 10-30 nuts for egg deposition; however, Criswell *et al.* (1975) found that females deposited eggs in an average of 7.8, 9.6, and 8.7 nuts of 'Hardy Giant', 'Mahan', and 'Nugget' varieties, respectively. In this same study, no differences were found in eggs per infested nut regardless of variety with an average of 4-4.5 eggs per pecan. Also, an average of 36.6 eggs/female was deposited in 'Hardy Giant' nuts compared with 53.8 in 'Mahan' and 44.4 in 'Nugget'. Harp and Van Cleave (1976a) had reported similar results for the number of eggs oviposited per female with averages of 34.7 and 54.7 for two years of study.

Recent work of Harris (1976b) indicates that the number of weevil larvae per infested pecan is apparently independent of the level of infestation, nut volume, nut weight or kernel weight. He also suggests that the independence of the number of weevil larvae per infested nut compared to the percent infestation on a tree support the view (proposed by Moznette *et al.* 1931) that previously infested nuts are avoided for subsequent oviposition.

Adult feeding. The amount and type of feeding damage by adult pecan weevils has a direct relationship to the population density, time of emergence, and longevity of weevils (Van Cleave and Harp 1971). Although it is generally known that pecans damaged by feeding punctures prior to shell hardening will drop prematurely, data concerning the magnitude of this type of damage is scarce.

Calcote (1975) reported that individual male and female weevils destroyed an average 0.23 and 0.29 nuts per day, respectively, early in the season (July 31-August 18) prior to shell hardening. The majority of the nuts punctured prior

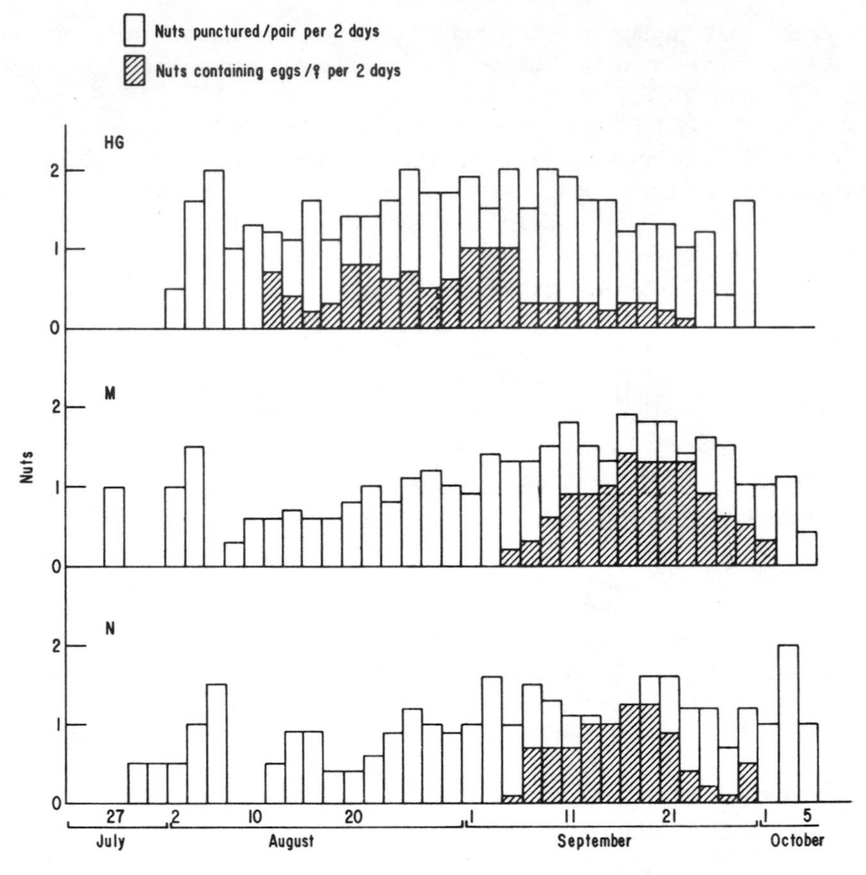

Figure 5. The average number of nuts punctured per weevils pair per 2 days and average number of nuts that contained eggs per female per 2 days on the pecan cultivars (Hardy Giant = HS; Mahan = M; Nugget = N) in the laboratory. [Taken from: Criswell *et al.* (1975)].

to August 18 fell in 6-15 days; whereas, nuts punctured after this date remained green and adhered to the tree past normal harvest (Table 1). After shell hardening, the males continued to feed, but only penetrated through the shucks. The nuts affected in this manner had slight scars on the shells but were undamaged. However, punctures by females after shell hardening had two effects: (1) if punctures penetrated through the cotyledonus layer (during the gel stage of nut development), molds readily deteriorated the gel resulting in "sticktight" or "pop" nuts at harvest; and (2) if punctures penetrated only into the cotyledonus layer, black spots, pits or molds occurred on the kernels near

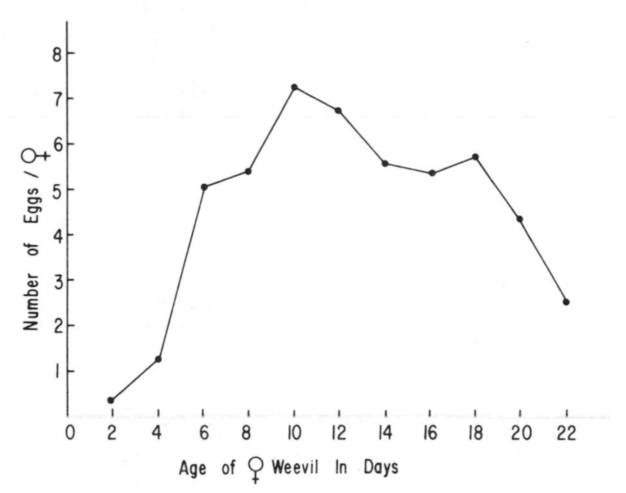

Figure 6. Ovipositional rhythm for 25 female pecan weevils living 22 days or longer. Taken from: Criswell *et al.* (1975).

the punctures. The latter damage resulted in a loss of 10-50% of the edible kernels.

Post emergence longevity. Van Cleave and Harp (1971) reported that post emergence longevity for adult pecan weevils averaged 8.6 and 17.6 days for males and 14.6 days and 23.1 days for females in studies conducted in 1967 and 1969, respectively. In both studies, the average longevity was 6 days greater for females than males. They concluded that those females which emerge early in the season are capable of surviving a sufficient length of time to find mature nuts in which to oviposit.

Criswell *et al.* (1975) found that weevils that emerged earlier in the season had a greater life span than those that emerged later in the season (Table 2). Their data also indicated that females are capable of surviving until nuts become acceptable for oviposition. Since the early emerging weevils tend to live longer, they have the capacity to inflict more damage than those weevils emerging later in the year.

Egg stage. Only empirical observations existed concerning the duration of the pecan weevil egg stage until Harp (1970) removed newly laid eggs from nuts and placed them on artificial diet to determine the time required for each egg to hatch. He found the duration to be from 6 to 14 days with an average of 9.8 days which was in close agreement with earlier literature reports.

Larval stage. Harp (1970) successfully reared weevil larvae on a cottonseed-meal diet (Sterling *et al.* 1965) in the laboratory and found four instars. Measurement of head capsules of laboratory-reared larvae closely correlated with

Table 1.

Effect of feeding and oviposition on Stuart variety pecans by pecan weevils confined to individual nut clusters, 1972. [a]

Weevils	Avg. no. [b] weevils present/day	No. nuts exposed	Total No. nuts punctured	No. nuts punctured through shuck only (undamaged)	Through-shell punctured nuts	% falling indicated days after puncture		
						6-15	16-38	38 +
Males	8.2	407	89	48	41	75.6	17.1	7.3
Females	13.3	643	159	65	94	60.6	12.8	26.6
Pairs	6.2♂, 10.8♀	425	167	51	85 (31[c])	65.9	9.4	24.7

[a] Taken from: Calcote (1975).

[b] Since some of the caged weevils died, this column reflects the average number of living weevils active each day.

[c] These 31 nuts were oviposited in; they were not used to calculate the percentages falling on given days.

Table 2.
Average longevity in days for pecan weevils emerging during the 3 periods of the season, 1972. [a]

Season	Cultivar								
	Hardy Giant			Mahan			Nugget		
	♀	♂		♀	♂		♀	♂	
Early (July 21–Aug. 8)	42.7	17.2	(4)[b]	56.0	40.0	(3)	33.7	37.2	(3)
Middle (Aug. 12–26)	24.4	13.2	(12)	35.6	24.3	(15)	21.6	22.8	(13)
Late (Sept. 1–17)	13.4	8.5	(5)	17.8	9.1	(6)	14.1	10.5	(6)

[a] Taken from: Criswell et al. (1975).
[b] Values in parentheses indicate the number of weevils of each sex observed to obtain the average.

those of field-collected larvae indicating the larvae undergo four instars in the field. The duration of the first three larval stages in the laboratory averaged approximately 13 days. Although the 4th instar larvae fed on the diet for 5 to 9 days, this instar spends from one to two years in the soil, without feeding, prior to pupation (Harp 1970).

Larval emergence. Upon completion of feeding inside the nuts, the larvae chew "exit" holes about one-eighth inch in diameter in the shell and emerge from the nuts and drop to the ground. The larvae burrow into the soil various depths depending on the soil type and condition of the orchard floor. Moznette *et al.* (1931) reported the larvae may penetrate as deep as nine inches but stated that most are found about six inches below the surface. Hinrichs and Thomson (1955) observed that larvae were found deeper in cultivated soil than in uncultivated soil. Weevil larvae show little lateral movement before or during the soil penetration (Chau 1949).

Early studies indicated that larvae emerge from the nuts during September, October, and November (Gill 1924; Leiby 1925; Bissell 1931). These reports were characterized by inconsistencies in the dates for onset of emergence and median emergence.

Recent investigations on native pecans (Harris 1976a) and improved varieties (Boethel and Eikenbary 1979) have demonstrated that emergence commenced earlier from the earlier maturing varieties or in the case of native pecans, the trees that matured their pecans the earliest. Assuming egg and larval development within the nuts to be similar regardless of variety, it is understandable that larval emergence would occur earlier on the early developing varieties that were susceptible to oviposition earlier in the season.

After the larvae leave the trees and burrow into the soil to the desired depth, they construct earthen cells in which they remain in diapause until pupation the following season.

Pupal stage. Harp and Van Cleave (1976a) reported that larvae placed in the soil in the fall of 1967 did not pupate until August 1968. About 10% of the larvae failed to pupate until the following season (August 1969).

Their field investigations revealed that adult and larval weevils could be found in the soil throughout the year; however, pupae were found only from mid September through the first week in October.

Laboratory studies demonstrated that the duration of the pupal stage averaged 19 and 18 days for males and 20 and 28 days for females in 1966 and 1967, respectively (Harp and Van Cleave 1976a). These data corresponded closely with field observations.

From these experiments, Harp and Van Cleave concluded that a majority of the pecan weevil pupate approximately a year after entering the soil with about 10% delaying pupation until the second year. The pupal stage lasted about three weeks.

Adult stage (subterranean portion). Harp and Van Cleave (1976b) noted that after transformation from the pupal stage the adults remained inactive within their earthen cells and did not attempt emergence until August of the following year (approximately 2 years after entering the soil as larvae). Their studies on the life history revealed that little mortality occurred in the subterranean larvae and pupae; however, subterranean adult mortality was considerable. Many dead adults were found in their earthen cells but the majority were within three inches of the soil surface with heads pointed upward. Thus it appears the most critical point in the pecan weevil life cycle involves its emergence from the soil.

Sex ratio. Adult emergence records from Georgia (Dupree and Bissell 1965) and field and laboratory investigations in Texas (Harp and Van Cleave 1976a) indicate a 1:1 sex ratio in the species. Knowledge of the sex ratio may become important in management of the pest especially as it relates to the utilization of phermones and will be discussed in more detail later.

Diapause. Because the pecan weevil spends approximately a year in the soil as a quiescent larva and another year as an inactive adult, it was generally accepted that these stages were in diapause. Harp and Van Cleave (1976c) conducted micro-respiration studies on the various developmental stages and found an 8-fold average increase in oxygen consumption in the active as compared to the inactive stages (postfeeding 4th instar larvae and preproductive adults) found in the soil. Further evidence to confirm diapause was the fact that the inactive stages had much greater fat body content than did the active stages, and the reproductive organs of the inactive stages were not fully developed.

The extended period of diapause involving two developmental stages has been the limiting factor in devising a laboratory rearing method for the pecan weevil. At present, researchers must depend on field collected specimens which are frequently difficult to obtain in desired quantities at the appropriate time. Both basic and applied investigations would be facilitated if the pest could be colonized and produced in the laboratory thus ensuring a constant supply of weevils for study.

Summary. In order to summarize the pecan weevil life history, the readers are referred to Fig. 2. Beginning in year 1, adult weevils emerge from the soil and mate during August and September. Both sexes feed on the pecans and after the shells become hardened and kernel formation begins, successful oviposition commences. The eggs and developing larvae remain in the nuts approximately a month. Upon completion of development, the 4th instar larvae chew "exit" holes in the pecans and drop to the soil during October and November. The larvae burrow into the soil and upon reaching the desired depth, construct earthen cells. They remain in these cells in a state of diapause until the following September (year 2) when pupation occurs. The duration of the pupal stage is about three weeks. After eclosion, the newly formed adults remain in the soil in diapause until August and September of the following season (year 3). Approximately 10% of the population undergo a two-year larval diapause which extends the life cycle an additional year.

The information concerning the two to three year life history of the pecan weevil and the various associated biological events was presented to aid in understanding the different approaches taken in the past and presently being developed for management of the pest. Hopefully, these background data will complement the discussions which follow.

HISTORICAL, CONTEMPORARY, AND PROSPECTIVE MANAGEMENT TACTICS

Cultural Control

Before the advent of effective insecticides, several cultural practices were investigated for control of pecan weevils. Probably the most widely practiced procedure was the removal of alternate hosts (hickory trees) from the vicinities of pecan orchards. Although other methods proved to be occasionally successful, their implementation diminished once adequate insecticides were found and hand labor became increasingly expensive and difficult to obtain.

Jarring. Early attempts to control pecan weevils involved jarring the limbs with padded poles or other devices to dislodge the adults. The weevils that fell were collected from sheets spread beneath the trees and subsequently destroyed. Swingle (1935) obtained an average reduction in damage of about 60% using this technique. The jarring commenced the middle of August and was continued at weekly intervals until the second week of September. Even though control was achieved in the study, concern was expressed about the number of sound immature nuts which might be dislodged by the procedure.

Jarring eventually evolved into a sampling method; however, even in this capacity, it did not receive widespread acceptance. The main reason was the inconvenience and impracticality of the technique for large commercial orchards (Neel and Shepard 1976).

Alteration of the orchard floor. After it became evident that the pecan weevil spent a considerable portion of its life history in the soil, it was only logical that attempts to alter this habitat dominated early control recommendations.

Gill (1924) stressed cultivation to destroy the larvae and pupae. Moznette *et al.* (1931) recommended cultivation in the winter or early spring followed by the confinement of chickens or hogs in the orchard to feed upon the insects turned up by the plow. They felt cultivation would be effective for adults and pupae but not the larvae which would burrow down again.

Bissell (1931) investigated another method of altering the orchard floor to prevent entrance of weevil larvae into the soil. He found that by removing the vegetation from the soil surface and compacting the ground with a heavy roller, successful larval penetration was reduced compared to plots where the surface was left undisturbed or only raked. A later experiment comparing larval penetration in cultivated and uncultivated soils proved to be negative, and it was determined that altering the soil surface could not be depended on to control larvae (Bissell 1934).

Although cultivation is still occasionally conducted in pecan orchards for weed control, rising fuel costs and the development of herbicides has reduced the practice. Probably more important is the fact that deep cultivation, which would be necessary to reach pecan weevils, would result in pruning of pecan feeder roots which are found near the soil surface.

Flooding. Attempts to control pecan weevil larvae by flooding the orchard floor resulted in a 33% reduction in population; however, this degree of control was not considered sufficient to make the technique practical (Nickels and Pierce 1947).

Physical Barriers. Gray *et al.* (1975) conducted several experiments to evaluate polyvinyl acetate (PVA) as a physical barrier to prevent weevil larvae from entering the soil. The results indicated that when PVA was applied to smoothed, bare ground, it did provide an effective barrier but when the orchard floor was disturbed or contained vegetation, the effectiveness declined. For this reason plus the high cost of the material, interest in the technique has subsided.

Biological Control

Pathogens. Among the first natural enemies reported for the pecan weevil were the fungi, *Metarrhizium anisopliae* (Metch.) and *Beauveria bassiana* (Bals.) (Swingle and Seal 1931). Recently, interest in these organisms as control agents has been renewed with laboratory studies yielding rather promising results.

Neel and Sikorowski (1972) found that relatively small amounts of innoculum of *B. bassiana* would cause mortality of larvae and adults. Depending on spore concentration, average adult mortality ranged from 38-77% and average larval mortality ranged from 70-75% with 7 and 10 days exposure, respectively. Tedders *et al.* (1973) studied various isolates of *B. bassiana* and *M. anisopliae* and observed that the effectiveness of the pathogens varied with isolate, culture technique, and method of application. However, 84% larval mortality was achieved in one test with a *B. bassiana* isolate. In field tests, maximum larval mortality obtained with *B. bassiana* was 61.5% while maximum mortality with *M. anisopliae* was 59.3%.

Monthly surveys for pecan weevil pathogens have been conducted in Mississippi and Arkansas (Sri-Arunotai *et al.* 1975). *B. bassiana* and *M. anisopliae* were found in all survey orchards with the latter species killing the highest number of larvae in Mississippi (10% in May collection) and Arkansas (23.6% in April collection). Two bacterial pathogens, *Pseudomonas aeruginosa* (Schroeter) and *Serratia marcescens* Bizio caused a maximum of 4.2% larval mortality in Mississippi (July) and 12.4% in Arkansas (September), respectively. Another fungus, *Synnematium* sp. also has been found attacking larvae in the field (Harp and Van Cleave 1976d).

Although field surveys have indicated that *B. bassiana* and *M. anisopliae* contribute to substantial larval mortality and laboratory tests with these fungi have been encouraging, few trials for augmentation in the field have occurred. It appears application of these entomogenous fungi to the air-soil interface during October and November would be a viable step toward integrated control of the pest. Applications at this point in the season when a majority of the larvae are entering the soil (Boethel and Eikenbary 1979) should offer the best opportunity for host infection and would occur after fungicide applications for pecan disease control have been curtailed.

Should entomogenous fungi become a management tool for pecan weevil, interdisciplinary cooperation will be essential to find pecan fungicides which are not detrimental to these pathogens. This will be particularly important in the humid southeastern section of the pecan belt where numerous fungicide applications (8-12 annually) are routine for disease suppression. Perhaps, runoff of currently recommended fungicides are presently limiting the impact of the indigenous fungi, *B. bassiana* and *M. anisopliae.*

Nematodes. Harp and Van Cleave (1976d) observed a nematode, *Neoaplectana* sp., parasitizing both 4th instar larvae and pupae of the pecan weevil destroying about 20% of a cohort of laboratory specimens. Tedders *et al.* (1973) found that *Neoaplectana dutkyi* Jackson (DD-136) caused greater weevil larvae mortality (67%) than three fungi in limited field trials. This nematode is currently the subject of biological control experiments being conducted in Florida. Preliminary results indicate that the pecan fungicides, benomyl and DuTer (triphenyl tin hydroxide), reduce nematode infectivity (Ball 1978).

Parasites. Swingle (1931) reportedly reared three dipterous parasites, *Myiophasia nigrifrens* Townsend, *Sarcophaga sima* Aldrich, and *Winthemia rufopicta* Bigot, from pecan weevil larvae. The parasitism attributed to these species was found to be less than 2%.

Harp and Van Cleave (1976d) found that a tachnid, *Myiophasia harpi* Reinhard, parasitized *ca.* 4% of a group of 4th instar larvae. Apparently, the life cycle of the fly coincides with that of the weevil larvae because some emerged as adults one year after entering their host while others emerged the following year. The parasites emerged from the soil during August and September, simultaneously with adult weevil emergence. Therefore, the adult flies appear to be active in the orchards when insecticide sprays are being directed at the adult weevils. These applications may be detrimental to the parasites thereby reducing their control potential.

Pheromones

After preliminary experiments indicated that the female pecan weevil may produce an aggregating pheromone (Van Cleave and Harp 1971), several laboratory experiments to identify the pheromone (Mody *et al.* 1973) and components of pecan leaves and nuts which might be precursors of the pheromone (Mody *et al.* 1976) have been conducted.

At this point, a pheromone has not been identified and field bioassays utilizing volatiles isolated from weevils have attained only limited success. Mody *et al.* (1973) reported that the distillate of males was more attractive to females and the distillate of females was more attractive to males. In the same tests, live females attracted more insects than did males.

Polles *et al.* (1977) also reported that female pecan weevils appeared to release a pheromone to which both sexes responded but their studies demonstrated that traps baited with grandlure (the boll weevil, *Anthonomus grandis* Boheman, pheromone) captured far more male weevils than live baited traps. The success with grandlure, which occurred during the 1972 growing season, prompted rather wide scale testing with the material throughout the pecan belt in 1973 with negative results. Apparently, a change in the components of the pheromone was responsible for the poor results and interest in grandlure has diminished.

Although unbaited traps captured as many weevils as those baited with grandlure in Oklahoma in 1973, some interesting results occurred concerning trap placement. Grovenburg *et al.* (1974) found that twice as many weevils were captured in traps suspended in the lower canopy of the trees and twice as many females as males were captured in the higher traps. Polles *et al.* (1977) mentioned that traps placed in the tree canopy captured more weevils than those placed on stands under the trees. Therefore, placement of traps can have effects on the number and sex of weevils caught and should receive consideration when interpreting results of pheromone field trials.

Other insect species are known to exhibit protandry (initial emergence of males earlier than that of females) which can be an important factor in the utilization of pheromones in an insect management program (Roelofs *et al.* 1970). Seasonal shifts in the sex ratio can alter the effectiveness of synthetic pheromones due to competition with natural populations.

Most reports on pecan weevil indicate a 1:1 sex ratio (discussed earlier) but no reports exist concerning seasonal shifts in sex ratio. Boethel *et al.* (1976b) used several sampling techniques to monitor adult emergence in Oklahoma. By combining the number of weevils captured by all methods and plotting the seasonal abundance of each sex, it appears that pecan weevils do not exhibit protandry and that the sexes occur in approximately equal numbers throughout the season (Fig. 7 and 8). The ramifications presented by pheromone trap placement and a seasonal sex ratio approaching unity may be clarified when it is determined how a pecan weevil pheromone will be utilized in management of the pest.

Host Plant Resistance

At the present time, there are no known seedlings or pecan varieties resistant to pecan weevil. There have been reports of certain varieties being more susceptible to attack.

Gill (1917) stated that the severest weevil infestations occurred on native pecans in Texas. He also reported that in an orchard in Georgia, trees of the 'Stuart' variety suffered more damage than those of 'Schley.' Moznette *et al.* (1931) found 'Schley' and 'Stuart' to be the most subject to attack evidently because they were among the earliest varieties to form kernels. A later publication added 'Rome' to the list of susceptible varieties (Moznette *et al.* 1940). Osburn *et al.* (1966) indicated that early maturing varieties such as 'Stuart', 'Schley', 'Mahan' and 'Moneymaker' were most commonly infested. Late maturing varieties such as 'Success', 'Teche', 'Mobile', and 'Van Deman' were not attacked unless the crop on the early varieties was light or had been destroyed by other pests before pecan weevils had finished feeding and ovipositing. All early reports indicated native hickory trees were severely infested.

The central theme of these published observations concerning varietal resistance to pecan weevil is that early maturing varieties suffer the most damage. In all probability, the conclusions drawn were based on the amount of damage evident at harvest which was the result of successful oviposition. No data exists concerning the susceptibility of varieties to adult feeding damage in early season prior to kernel formation. It would be interesting to know whether the degree of feeding damage differs among varieties of differing phenology.

As mentioned previously, the early maturing varieties are acceptable for oviposition over a longer time span. In years when weevil emergence occurs early and is completed quickly, late maturing varieties may escape damage. This

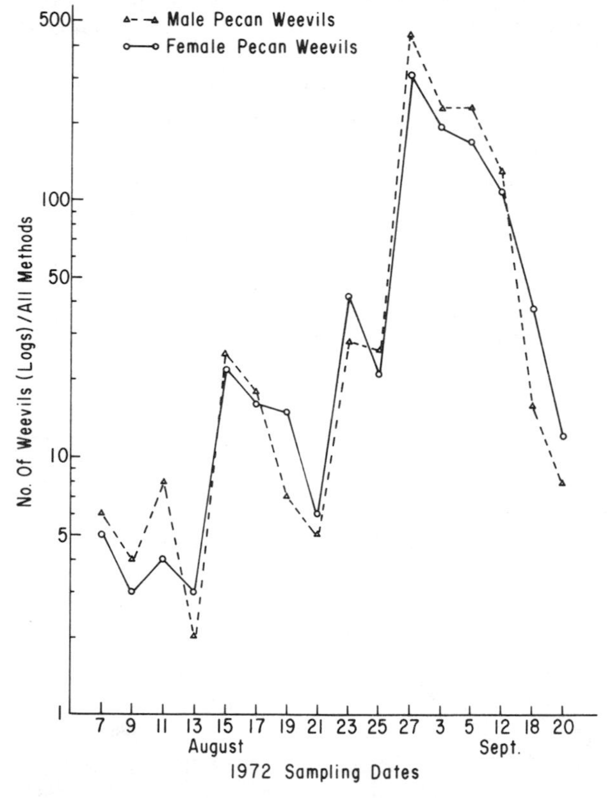

Figure 7. Seasonal emergence of adult pecan weevils. Data obtained by four sampling methods. Stillwater, Oklahoma, 1972.

explanation for the decreased damage on late varieties is questionable when research has shown the pest usually has an extended adult emergence period. Also, the post emergence longevity of the adults indicates they are capable of surviving to oviposit and damage late maturing pecans. The observations of Osburn *et al.* (1966) confirm these statements.

In general, improved pecan varieties produce larger yields than native pecans but due to their larger size, produce fewer nuts per acre. Since the pecan weevil apparently does not discriminate among pecans of varying sizes, about equal numbers of large improved pecans will be infested, compared to smaller native pecans under the same weevil infestation conditions. However, the greater loss in yield would occur in the improved pecans from a given weevil infestation. Thus it appears that development of pecan varieties that yield more nuts per acre would be desirable to reduce yield losses to pecan weevil (Harris 1976b).

Development of pecan varieties is a long term process. Brison (1974) listed 15 varieties which have been released since 1930 and the average lapse of time from date of pollination until the varieties were introduced was 22.8 years. Pecan varieties are predominately selected on desirable horticultural characteristics (precocious and prolific) with little consideration being given to

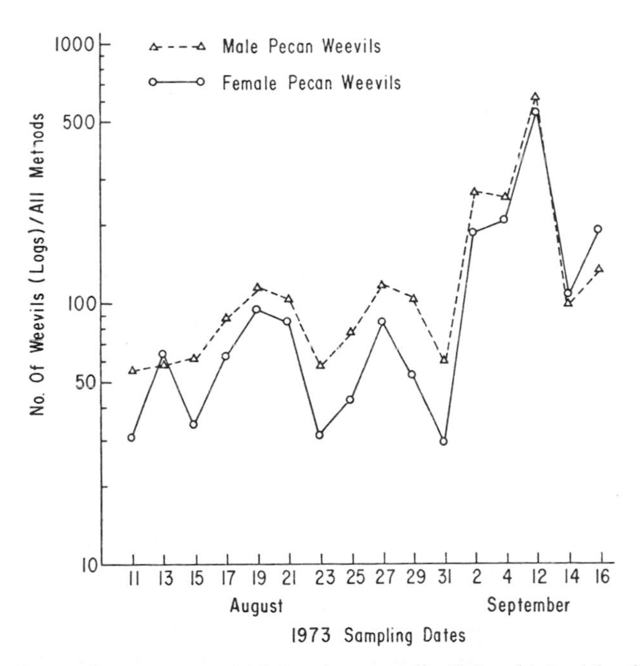

Figure 8. Seasonal emergence of adult pecan weevils. Data obtained by four sampling methods. Stillwater, Oklahoma, 1973.

traits such as insect resistance. To test varieties for their tolerance to insects would increase the selection process further and perhaps, is the reason it has not been incorporated already. Since some new selections produce nuts as early as the 4th year rather than 10-12 years as in the past, time may no longer be an obstacle.

Chemical Control

Similar to the situation with other insects of economic importance, pecan weevil literature is dominated by reports on insecticide screening trials to find effective materials to control the pest. To cite the specifics of the various tests would result in a great deal of repetition; therefore, only highlights will be given for an understanding of the evolution of the present day insecticide control programs.

Early insecticide control (1900-1960). The first attempts to utilize chemicals to control the pecan weevil involved early harvest of the pecans followed by fumigation with carbon disulphide to kill the larvae inside the nuts (Gill 1924; Leiby 1925). It was thought this approach would reduce the weevil problem in subsequent years.

During the two decades following World War II, numerous field trials were conducted to find effective foliar pesticides. Although lead arsenate, benzene hexachloride, dieldrin, toxaphene, and parathion occasionally provided some control of small populations, the consensus was that DDT was the most effective compound particularly when small to moderate populations were encountered (Moznette 1947; Hinrichs 1948; Nickels 1950; Osburn 1952; Dupree and Beckham 1953; Hinrichs and Thomson 1955). When large infestations occurred, even DDT was ineffective (Nickels 1950; Osburn 1952). All investigators indicated a need for multiple applications.

During this era attempts to kill the insect in the soil with chemicals gave unsatisfactory results (Hinrichs 1948; Nickels 1952).

Contemporary insecticide control (1960-present). The search for effective pecan weevil insecticides continued during the 1960's. At the beginning of the decade, DDT was the material of choice but due to its aggravation of aphid and mite problems, toxaphene was recommended also. Toxaphene presented difficulties because of its incompatibility with Bordeaux mixture which was still being used in some localities for disease control (Osburn *et al.* 1963). EPN was reportedly capable of controlling light and medium weevil populations, and a later publication reported that low volume sprays of EPN were as effective as the conventional dilute treatments (Osburn *et al.* 1964).

Osburn and Tedders (1966) found carbaryl and Du-Ter to be effective for control of pecan weevil but noted that on trees treated with these compounds the black pecan aphid, *Tinocallis caryaefoliae* (Davis) was more prevalent than on check trees that only received the fungicide, dodine. Since adult pecan weevil emergence coincides fairly closely with treatments for hickory shuckworm, *Laspeyresia caryana* (Fitch), control, Van Cleave *et al.* (1969) attempted to find insecticides that would control both pests. They reported that multiple applications of EPN and azinphosmethyl effectively reduced heavy weevil infestations.

Even though early attempts to control the pecan weevil in the soil with pesticides were unsuccessful, renewed interest in this approach occurred during the 1970's. Laboratory and field trials demonstrated varying amounts of control (Tedders and Osburn 1971; Polles *et al.* 1973; Payne *et al.* 1975; Gentry *et al.* 1977; Neel *et al.* 1976; Pollet and Aitken 1978). With the current knowledge concerning inter-tree dispersal of pecan weevil adults, it appears soil treatments will have to be supplemented with foliar sprays in order to protect the nuts from weevils moving into the canopies from untreated trees.

Because rainfall in late season occasionally precludes the use of ground sprayers and the fact that the large air-blast sprayers are prohibitively expensive for growers with small acreage, interest in aerial application has surfaced from time to time. The results obtained by the method have not been encouraging. Pierce (1963) reported negative results with parathion in tests in Louisiana. Ellis and Polles (1976) found that several materials when applied by air reduced weevil damage; however, they indicated further testing was needed to draw definite conclusions.

In recent years, extensive field screening trials to find effective foliar pesticides have taken place in various states across the pecan belt: Georgia (Polles and Payne 1974; Payne *et al.* 1975; Tedders 1976; Polles 1976), Mississippi (Neel *et al.* 1976), Louisiana (Calcote 1974), and Texas (Harris and Aguirre 1978). In all these tests, carbaryl proved to be the most effective material and is now considered the standard insecticide for control of the pecan weevil.

Carbaryl has some of the limitations that beset DDT. Since multiple applications of the insecticide (2 to 5 per season) have become routine, concern about build-up of mites and aphids has resulted in the addition of an aphicide or miticide to one or more of the carbaryl applications. This management approach has increased the total pesticide load in pecan orchards and has increased production costs.

For this reason, an effort has been made to find insecticides that will control the pecan weevil but not create further problems with aphids and mites, either through direct kill of the species or through reduced impact on the natural enemies of these foliage pests. Dialifor and phosalone have demonstrated sufficient weevil control in field trials to obtain registration. These materials have limitations, too. To achieve control comparable to carbaryl, it appears dialifor must be applied more frequently and at shorter intervals (Payne *et al.* 1976). Also during the 1978 growing season, some difficulty was experienced in controlling aphid outbreaks with dialifor in Louisiana (Boethel, personal observation). Only infrequently has phosalone exhibited pecan weevil control comparable to carbaryl, and its registration label currently indicates it will provide "weevil suppression."

Both dialifor and phosalone usually provide better control of hickory shuckworm, another late season pest, than does carbaryl. However, given a choice between effective control of shuckworm or pecan weevil, the grower is prudent to choose the latter.

Some southeastern states have continued to include EPN, azinphosmethyl, and toxaphene in their state recommendations (Anon. 1978); therefore, on paper it appears that a rather substantial arsenal of insecticides is available for pecan weevil control. This assumption is erroneous because none of the compounds mentioned have given the consistent control of carbaryl. In actual practice, pecan weevil management has become rather dependent on this insecticide.

Since the pecan weevil is a univoltine insect with a two to three year life cycle, the possibility of it developing resistance in the near future appears remote. With a program relying solely on one insecticide resistance should not be taken lightly, however. Polles and Payne (1973) evaluated several pesticides in the laboratory, using the nut-dip and residual-plate methods, against field-collected pecan weevils. This work constitutes the only base-line toxicological data gathered on the pest. Undoubtedly, the lack of specimens, resulting from the inability to rear the insect in the lab, has hampered this type of research. The precarious nature of a management program dominated by a single insecticide demands additional laboratory and field research to find alternative pesticides.

DEVELOPMENT OF A PEST MANAGEMENT PROGRAM

Sampling

By the late 1960's, most pecan entomologists realized that many of the control failures experienced in the past could be attributed to improper timing of spray applications. The sampling technique most widely used (jarring) was unsatisfactory, and prophylatic treatments without regard to the populations present were commonplace. It became evident that new sampling methods were required to improve timing of insecticide applications and establish the actual need for multiple treatments. Therefore, the early years of the 1970's, were characterized by considerable interest in pecan weevil sampling technology, and marked the first step toward a pest management approach to the pecan weevil problem.

Neel and Shepard (1976) reviewed the "state of the art" of sampling for adult pecan weevils. The publication pointed out some merits and deficiencies of the various techniques but did not recommend any particular method. Boethel *et al.* (1976b) compared four sampling techniques: limb jarring, tanglefoot trunk barriers, cone emergence traps, and quick knockdown insecticide sprays. The results indicated that the emergence trap (Raney and Eikenbary 1969) was the most practical method to detect the onset of adult emergence, fluctuations in emergence, and peak emergence.

Because several researchers had observed that certain pecan trees in an orchard had large infestations of pecan weevils and that nearby trees were virtually uninfested, interest turned to the inter-tree variation in weevil populations and the consequence this might have on sampling procedures. Although the test trees were selected for uniformity in size, age, variety, and edaphic conditions, considerable variation in populations did occur among trees (Table 3). An attempt to estimate the number of trees to be sampled for a specified precision was then undertaken (Boethel *et al.* 1976b). The variation associated with the pecan weevil population and, consequently, each sampling technique, was greater in 1972 than 1973, resulting in more trees being estimated for 1972. As can be seen in Table 4, on the earlier sample dates when population levels were low, more trees were required; however, near peak emergence, the numbers declined. On most sample dates, the estimated number of trees to be sampled for a 10% standard error was too large to be practical but one-quarter as many trees would be needed if only a 20% standard error were required. Since the latter option revealed that near peak emergence a more realistic number of trees would be required and no data existed as to the degree

Table 3.

Total number of pecan weevils collected per tree by sampling techniques. Stillwater, Oklahoma, 1972-1973. [a]

	Tree Number									
Year [b]	1	2	3	4	5	6	7	8	9	10
1972	154	166	248	353	286	320	131	71	179	262
1973	216	227	273	610	430	744	361	283	405	553
Total	370	393	521	963	716	1064	492	354	584	815

[a] Taken from: Boethel et al. (1976b).

[b] Only data that was collected on two-day intervals (sampling dates) for each year were included.

of precision needed for sampling pecan weevil populations, it was decided that future sampling studies would concentrate on the emergence trap technique utilizing 10 trees with 12 traps per tree.

Prediction Equations

Prior to the work of Raney *et al.* (1970), little data existed concerning estimates of pecan weevil populations. Since the study comparing sampling methods indicated that the emergence traps might be the technique of choice for adult pecan weevils, data obtained by sampling studies during 1972 and 1973 (Boethel *et al.* 1976b) were used to develop prediction equations to estimate the pecan weevil population in a tree on a given date from the number of weevils collected from emergence traps beneath that tree.

The following least squares multiple regression was fitted to the data:

$$Y_{ij} = b_o + T_i + b_1 D_j + b_2 X_{ij} + b_3 D_j^2 + b_4 X_{ij}^2 + b_5 D_j X_{ij} + E_{ij}$$

where

Y_{ij} = the total number of pecan weevils collected from the ith tree on the jth sampling date

T_i = effect due to the ith tree

D_j = number of days from the 1st sampling date

X_{ij} = number of weevils collected from 12 emergence traps under the ith tree on the jth sampling date

E_{ij} = random error associated with Y_{ij}.

Thus, given the appropriate values for b's, the predicted number of pecan weevils ($\hat{Y}$) could be obtained for a given tree (T), with a trap value (X), and a sampling date (D). The b's in the foregoing equation were the least-squares solutions to the normal equations.

In the earlier sampling studies, it was observed that pecan weevils could be collected on the outside of emergence traps; therefore, some prediction equations were built from trap data which utilized the weevils collected both inside and outside of the emergence traps. In Table 5, the models A, A′, C and C′ represent equations built from trap data based on weevils captured both inside and outside the traps. The models B, B′, D, and D′ utilized trap data based only on weevils captured inside the traps.

Also in the earlier sampling studies, it was observed that later in the pecan weevil season after peak emergence, movement of pecan weevils occurred into the test trees from the border trees. Due to this phenomenon, it was felt that estimates based on trap data late in the season might be biased; therefore, prediction equations were developed from data collected early in the season presumably prior to any dispersal. To separate the sampling dates as to early season or late season and to avoid bias on the part of the experimenter in making this separation, arbitrary criteria were used to separate the early sampling dates as follows: all dates were included after the date one weevil was trapped under any tree and all dates were excluded after the date 10 or more weevils were trapped under any tree. In Table 5, the models A, B, C, and D represent equations built from early season data, and models A′, B′, C′, and D′ represent equations built from data collected over the entire sampling season.

Table 4.

Number of pecan trees to be sampled for a 10%[b] standard deviation of the mean in estimating populations of the pecan weevil. Stillwater, Okla., 1972 and 1973[a]

Sample date	Emergence traps				Spray				All[c]			
	C.V.		No. trees		C.V.		No. trees		C.V.		No. trees	
	'72	'73	'72	'73	'72	'73	'72	'73	'72	'73	'72	'73
1	108	99	116	98	218	62	474	39	98	62	95	39
2	218	71	474	50	126	56	157	31	99	54	98	30
3	231	66	532	44	108	70	116	49	104	58	108	33
4	173	80	298	64	218	54	474	29	156	52	242	27
5	98	73	95	53	81	44	66	19	77	39	59	15
6	217	80	474	64	62	37	39	14	60	36	37	13
7	156	94	242	89	101	75	102	56	83	69	69	47
8	218	98	474	95	130	60	168	36	106	55	112	30
9	88	122	78	149	76	49	57	24	64	48	41	24
10	123	108	154	116	64	39	41	16	49	39	24	15
11	45	99	21	98	54	56	29	33	48	57	23	33
12	54	68	29	46	29	59	8	35	28	59	8	35
13	64	69	41	48	36	52	13	27	35	53	12	28
14	71	59	50	35	37	23	14	5	35	24	12	6
15	316	62	1000	39	47	40	23	16	47	37	22	14
16	316	78	1000	60	66	37	44	14	65	35	42	12

[a] Taken from: Boethel et al. (1976b).

[b] To obtain the number of trees to be sampled for a 20% standard deviation of the mean, divide the number of trees by four.

[c] "All" represents the sum of all weevils obtained by the methods used each year.

Table 5.

Prediction equations for estimating numbers of pecan weevils. Stillwater, Oklahoma, 1972-1973. [a]

Model [b]	Season	Trap Data	b_0	b_1	b_2	b_3	b_4	b_5	R^2
					1972				
A	Early [c]	Inside + Outside	-0.960	0.310	0.308	-0.009	-0.001	0.090	0.94
A'	Entire	Inside + Outside	-2.000	0.499	-7.141	-0.006	-0.017	0.382	0.87
B	Early [c]	Inside	1.650	-0.244	-10.778	0.017	-0.306	0.863	0.78
B'	Entire	Inside	-6.100	0.933	-3.062	-0.011	-0.053	0.449	0.72
					1973				
C	Early [d]	Inside + Outside	9.083	-0.207	-0.354	0.016	0.134	0.211	0.75
C'	Entire	Inside + Outside	11.568	-0.399	-0.218	0.016	-0.109	0.266	0.71
D	Early [d]	Inside	14.186	-1.050	-2.323	0.054	0.353	0.349	0.58
D'	Entire	Inside	13.286	-0.691	-1.333	0.036	-0.343	0.450	0.55

[a] Taken from: Boethel et al. (1976a).

[b] Fitted equations are of the form $Y_{ij} = b_o + T_i + b_1 D_j + b_2 X_{ij} + b_3 D_j^2 + b_4 X_{ij}^2 + b_5 D_j X_{ij}$.

[c] Early season data in 1972 included sampling dates August 3, 1972 - August 27, 1972.

[d] Early season data in 1973 included sampling dates August 11, 1973 - September 2, 1973.

Since model A and model C had the largest R^2 values for 1972 and 1973, respectively (Table 5), these models were evaluated to determine their effectiveness for predicting pecan weevil populations. When the population estimates obtained from the model built from early season data in 1972 (model A) were compared with the actual populations found in the trees in 1973, it was found that 71.7% of the observed values in 1973 fell within the 95% confidence intervals for the predicted values. Similar comparisons were made utilizing the model built from early season data in 1973 (model C) and 95% of the observed population values in 1972 fell within the confidence intervals for the predicted values.

When the confidence intervals were built for testing the effectiveness of the models, the effects due to the ith tree (T_i) were incorporated into the calculations. Thus by knowing which of the 10 experimental trees was being tested, the precision in predicting the weevil populations by these models was increased. In actual field situations, the values for T_i would be unavailable due to the lack of information about the inter-tree variation in populations in most orchards. The only option available was to use the models for the "average" pecan tree (T_i= O) but this alternative required caution because the sampling data indicated large tree to tree variation in pecan weevil populations. However, the preliminary data concerning population estimation based on emergence trap collections were encouraging and demonstrated the technique had potential for incorporation into a management program for the pest.

Modification and Validation of Prediction Equations

Because the prediction equations appeared promising for estimating pecan weevil populations, R.D. Eikenbary and co-workers in Oklahoma embarked on a series of experiments to validate the technique under field conditions. The original equations were formulated based on data derived from two-day intervals. Obviously, a model that only estimated populations over two days would have limited value in a management program unless the predicted values could be accumulated over time and still accurately estimate the population. In this case, natural survival would become a variable which was not considered when the original models were built.

In order to incorporate a factor for natural survival into the models, survival data taken from laboratory studies by Criswell *et al.* (1975) were used. When the original equations with minor changes in the interpretation and application of the terms of the equations were combined with the survival factor, new "modified" equations resulted which were utilized in subsequent studies (see Eikenbary *et al.* 1977 and 1978).

In 1974, validation experiments were conducted to measure the accuracy of the modified equations (model C) in predicting the number of weevils in the trees (Eikenbary *et al.* 1977). On 21, 31 August and 8, 14, 22 September, the traps were removed from beneath the trees and the entire area within the drip line of the 10 experimental trees was covered with 6 mil polyethylene tarp. The trees were then sprayed with a quick knockdown insecticide and approximately 30 minutes after spraying, the weevils were collected from the tarps, counted and recorded for each tree. After the tarps had been checked, they were removed and the emergence traps replaced in their original locations. This process was repeated on each spray date.

The results of the evaluation of model C are shown in Fig. 9. The equation slightly underestimated the average number of weevils in the trees on August 31, slightly overestimated on September 8, correctly estimated on September 14, and overestimated on September 22. When emergence was terminating at the end of the season, the equation was less accurate, but accuracy may not be as important at this time because weevil longevity, number of eggs deposited, and number of nuts damaged are greatly reduced (Criswell *et al.* 1975).

Economic or Action Threshold

To determine whether the prediction equations could be used in a pecan weevil management program as a tool for determining the need for insecticide treatments and the proper scheduling of the treatments, a vital element, the economic (action) threshold, was missing. Eikenbary *et al.* (1977) proposed that an average of 40 weevils per tree would constitute this threshold and derived it in the following manner. Criswell *et al.* (1975) reported that female weevils oviposited in 7.8-9.6 nuts. Therefore, if it took an average of 8.7 nuts for oviposition per female then 20 females (sex ratio of 1:1) would oviposit in 194 nuts, which would be about 3.5 lbs. if a variety averaged 50 nuts per pound. Utilizing this rationale, 3.5, 1.75, and 0.85% damage would occur on trees producing 100, 200, and 300 lbs. of pecans annually. Thus on trees averaging 200 lbs. of nuts per year, damage by oviposition would be below 3% which was considered acceptable.

Management of Populations With Prediction Equations

Simultaneously with the validation experiments, studies were undertaken to determine whether the prediction equations could be used to manage pecan weevil populations in an orchard known to have heavy weevil infestations. An experimental plot consisting of 10 trees was selected. The number and arrangement of traps, and the method of collecting and recording data were the same as described by Boethel *et al.* (1976b). When the average predicted number of weevils for the 10 trees exceeded 40 (economic threshold), the trees in the plot were sprayed with carbaryl (Sevin®). If inclement weather prevailed, the insecticide was applied at the earliest possible time after the predicted threshold was reached. To determine the longevity of carbaryl and know when to start

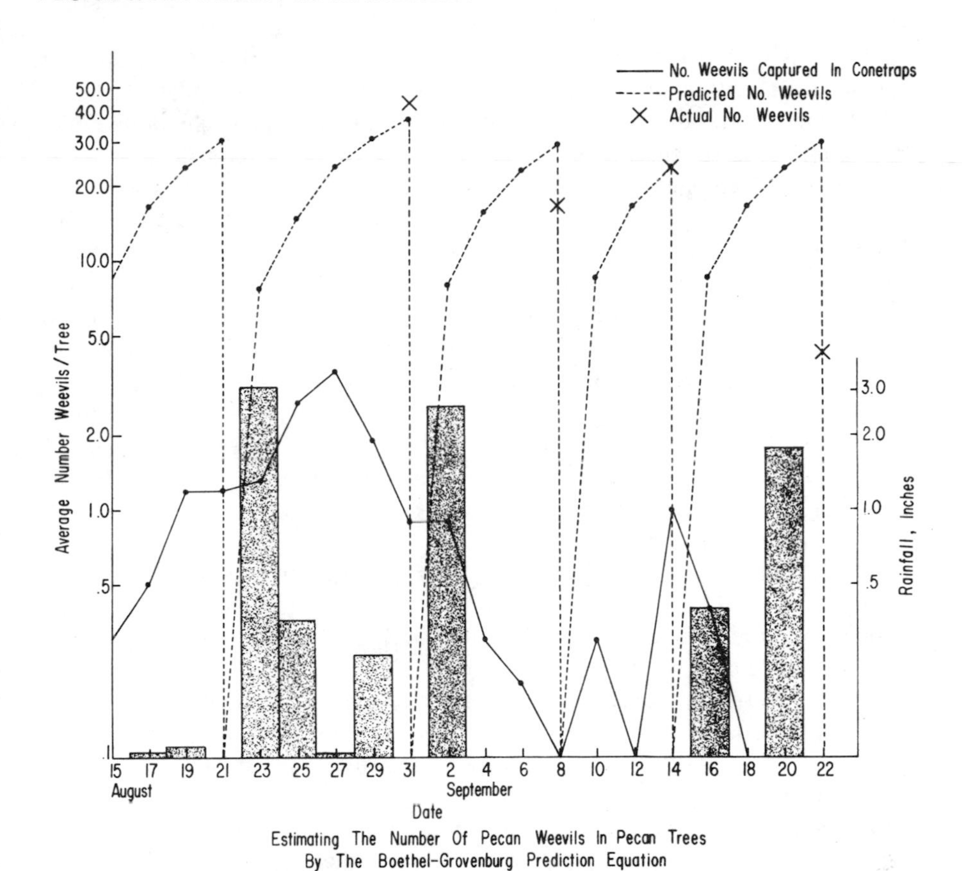

Figure 9. Relationship among the number of pecan weevils collected in cone traps, the estimated, and the actual number of weevils in the trees. Stillwater, Oklahoma, 1974. Taken from: Eikenbary *et al.* (1977).

recording emerging weevils for predicting weevils in the tree after the spray applications, the soil under each tree was checked daily for dead weevils. If trap observations showed weevil emergence, and no dead or dying weevils were found, the residual effect of the insecticide was considered terminated.

Timing of insecticide applications based upon a threshold of 40 weevils per tree as estimated by the predication equations and modified with the survival factor, was successful for three study years (Eikenbary *et al.* 1978). The treated trees had 2.4, 0.6, and 0.4% weevil damage compared to untreated trees which had 64.1, 56.9, and 70.4% damage in 1974, 1975, and 1976 respectively. The treated trees received three applications of carbaryl in 1974 and 1975 and two applications in 1976. Usually three or more applications are recommended for weevil control with conventional spray programs.

A total evaluation of a pecan weevil management program should consider the number of larvae that go into the soil to propagate the species. If perfect control were obtained one year, few or no weevils would be available 2-3 years hence to damage nuts and propagate the species. The high number of weevils trapped (> 350 under a tree) in 1974 and low numbers trapped (<30 under the same trees) in 1976 indicated management using the prediction equations might reduce the weevil populations over time. Since the experimental trees averaged 205 pounds of nuts in 1974, the scarcity of oviposition sites should not have been a factor in the reduced populations encountered in 1976.

These results indicating population reduction over time were interesting considering the control approaches utilized in some states to accomplish the same objective. Because of the link between oviposition and shell hardening, some states recommend initiation of insecticide applications when the shells begin to harden on the earliest maturing varieties. Subsequent applications are usually recommended at predetermined intervals until weevil activity is believed to have ceased. After 2-3 years of concentrating the spray program during the period when the nuts are susceptible to oviposition, the pecan weevil population in an orchard should be reduced.

Although this approach has merit, it has negative aspects, also. It does not take into consideration the populations present, and the preventive nature of the program may result in unnecessary treatments. As stated earlier, the development of resistance in pecan weevil appears remote; however, numerous applications against the weevil may induce secondary pest outbreaks and contribute to resistance development in other pests which appear simultaneously, such as aphids and mites. By delaying the initial application until shell hardening, feeding damage is ignored, and even though this damage will not contribute to the population two years hence, it will reduce yields in the current season.

In orchards with histories of severe infestations, a two year program of insecticide sprays timed on nut phenology could be supplemented by a management program utilizing emergence traps and prediction equations in the third and subsequent years. Thus after severe populations have been reduced, reductions in the number of applications may be possible. At least, the combination of these approaches would offer an alternative to continuation of the preventive approach.

Implementation

Before a pecan weevil management program based on population estimates by prediction equations could be implemented by pecan growers, several problems had to be resolved. First, to eliminate extensive calculations, tables were developed which listed trap catch values and the corresponding population estimates for those values. Second, to ease computation, simplified flow charts were developed for recording trap catch, estimated weevils in the trees, and for determining when the economic threshold had been reached. Third, since most conventional spray programs allowed 10 to 14 days between applications, it was decided that 7 days should pass after a spray application before the prediction process was resumed. However, if no weevils were collected in the emergence traps on day 8, the process still would not resume until the day the first weevil was captured.

With these refinements, the program has been implemented by pecan growers in Oklahoma and has proven to be a successful tool in weevil management (Hedger *et al.* 1978). Limited trials utilizing the technique have also been conducted by pecan researchers in Louisiana, Florida, Alabama, and Georgia. The latter two states have pilot pecan pest management programs sponsored by their respective cooperative extension services, and there are plans to include the technique in these programs during the 1979 growing season.

SUMMARY AND CONCLUSIONS

Obviously, using the traps and trap arrangement suggested by Boethel *et al.* (1976b) and the prediction equations as suggested by Hedger *et al.* (1978), the ultimate pecan weevil management system has not been developed. The method does allow for standardization of sampling for adult weevils and collection, storage, and retrieval of data. The technique has demonstrated promise for both short and long term pecan weevil management. As evaluation continues and the program extends into other geographic areas, data will be acquired and incorporated which should result in improvement of the existing program.

In their review of the status of integrated control of fruit pests, Hoyt and Burts (1974) indicated that more permanent solutions to pest problems of these ecosystems might be found if much of the research effort was directed toward a multi-faceted attack on a few key pests.

The pecan weevil is a key pest of pecans, thought to be the most damaging insect attacking the crop, and has been the subject of considerable research. For this reason, this chapter has been devoted exclusively to discussion of the development of pest management programs for this insect.

Throughout the chapter, the terminology "integrated pest management" has not been used. Although research is underway to find alternative control methods, it appears that in the near future management programs will continue to be dominated by insecticide control. This is certainly not a unique situation as many key pests are not readily controlled by methods other than pesticides.

However, through the use of monitoring techniques, economic thresholds, population estimates, etc., more judicious "management" of the pecan weevil has occurred. Further improvements in the pecan weevil program plus advances in management of other insect and disease pests (discussed in the next chapter) may result in an over-all reduction of the pesticide load in pecan orchard ecosystems.

ACKNOWLEDGMENTS

The authors are grateful to the following colleagues for their critical review and valuable suggestions concerning portions of this manuscript: J.M. McBride and R.S. Sanderlin, L.S.U. Pecan Research and Extension Station; R.D. Morrison, Oklahoma State University; M.K. Harris and H.W. Van Cleave, Texas A & M University; and J.A. Payne, USDA-SEA, Southeastern Fruit and Tree Nut Research Station.

REFERENCES

Anonymous. 1970. The pecan weevil: A major crisis facing the pecan industry. Unpublished report to the Federated Pecan Growers Association of the United States and The National Pecan Shellers and Processors Association.

Anonymous. 1977. Louisiana Livestock and Crop Reporting Serv. Alexandria, La.

Anonymous. 1978. Preparing for spring spraying. *Pecan South* **5(3)**: 124-134.

Ball, J. C. 1978. Annual report of cooperative regional project S-88. "Bionomics and Control of the Pecan Weevil." 3 pp.

Barry, R. M. 1974. Costs of insect damage. *Pecan South* **1(4)**: 33.

Bissell, T. L. 1931. Experiments on controlling larvae of the pecan weevil by cultural methods. *J. Econ. Entomol.* **24**: 861-866.

Bissell, T. L. 1934. Ineffectiveness of surface cultivation to prevent the burrowing of pecan weevil larvae into the soil under pecan trees. *J. Econ. Entomol.* **27**: 1128-30.

Boethel, D. J. 1974. Pecan weevil, *Curculio caryae* (Horn): Comparison of sampling techniques, estimation of populations, and determination of the relationships between certain physical parameters and adult emergence. Ph.D. Thesis. Oklahoma State University. Stillwater, Oklahoma.

Boethel, D. J. 1978. Unpublished data. Shreveport, La.

Boethel, D. J., and R. D. Eikenbary. 1979. Seasonal emergence of larvae from several pecan cultivars in Oklahoma. *J. Ga. Entomol. Soc.* **14**: (In Press).

Boethel, D. J., R. D. Morrison, and R. D. Eikenbary. 1976a. Pecan weevil, *Curculio caryae* (Coleoptera: Curculionidae) 2. Estimation of adult populations. *Can. Entomol.* **108**: 19-22.

Boethel, D. J., R. D. Eikenbary, J. R. Bolte, and C. R. Gentry. 1974. Sampling pecan weevil infestations: Effects of tree, height and sector. *Environ. Entomol.* **3**: 208-210.

Boethel, D. J., R. D. Eikenbary, R. D. Morrison, and J. T. Criswell. 1976b. Pecan weevil, *Curculio caryae* (Coleoptera: Curculionidae) 1. Comparison of adult sampling techniques. *Can. Entomol.* **108**: 11-18.

Brison, F. R. 1974. Pecan Culture. Capitol Printing, Austin, Texas. 292 pp.

Calcote, V. R. 1974. Combination of two insecticides and a sticker tested against the pecan weevil on pecan. *J. Econ. Entomol.* **67**: 695-696.

Calcote, V. R. 1975. Pecan weevil: Feeding and initial oviposition as related to nut development. *J. Econ. Entomol.* **68**: 4-6.

Chau, K. M. 1949. Study of feeding habits and investigating soil fumigation as a means of control for pecan weevil. M.S. Thesis. Oklahoma State University, Stillwater, Oklahoma.

Chittenden, F. H. 1908. The nut weevils. *USDA Circ.* 99. 15 pp.

Criswell, J. T., D. J. Boethel, R. D. Morrison, and R. D. Eikenbary. 1975. Longevity, puncturing of nuts, and ovipositional activities by the pecan weevil on three cultivars of pecans. *J. Econ. Entomol.* **68**: 173-177.

Dupree, M. and C. M. Beckham. 1953. A two-year study of insecticides for control of pecan weevil. *Proc. Southeastern Pecan Growers Assoc.* **46**: 95-98.

Dupree, M., and T. L. Bissell. 1965. Observations on the periodic emergence of the pecan weevil. *Proc. Southeastern Pecan Growers Assoc.* **58**: 50-51.

Eikenbary, R. D., and H. G. Raney. 1973. Intratree dispersal of the pecan weevil. *Environ. Entomol.* **2**: 927-930.

Eikenbary, R. D., W. G. Grovenburg, G. H. Hedger, and R. D. Morrison. 1977. Modification and further evaluation of an equation for predicting populations of *Curculio caryae* (Coleoptera: Curculionidae). *Can. Entomol.* **109**: 1159-1166.

Eikenbary, R. D., R. D. Morrison, G. H. Hedger, and D. B. Grovenburg. 1978. Development and validation of prediction equations for estimation and control of pecan weevil populations. *Environ. Entomol.* **7**: 113-120.

Ellis, H. C., and S. G. Polles. 1976. Aerial application for the control of certain pecan insects. *Proc. Southeastern Pecan Growers Assoc.* **69**: 117-121.

Gentry, C. R., J. S. Smith, R. E. Hunter, J. A. Payne, and J. M. Wells. 1977. The second year of an integrated program for pest management on pecans. *Proc. Southeastern Pecan Growers Assoc.* **70**: 155-163.

Gibson, L. P. 1969. The genus *Curculio* in the United States. *Entomol. Soc. Am. Misc. Pub.* **6(5)**: 239-285.

Gill, J. B. 1917. Important pecan insects and their control. *USDA Farmers Bull.* 843. 40 pp.

Gill, J. B. 1924. Important pecan insects and their control. *USDA Farmers Bull.* 843. 48 pp.

Gray, L. B., W. W. Neel, and J. A. Payne. 1975. The use of polyvinyl acetate as a barrier to the pecan weevil larvae. *Pecan South* **2(5)**: 194-196.

Grovenburg, W. G., D. J. Boethel, and R. D. Eikenbary. 1974. Attraction of the pecan weevil to the synthetic boll weevil pheromone. *Proc. Oklahoma Pecan Growers Assoc.* **44**: 72-76.

Harned, R. W. 1929. A summary of the report of pecan investigations for 1928-1929. *Miss. Agr. Exp. Sta. Ann. Rept.* **42**: 43.

Harp, S. J. 1970. The biology and control of the pecan weevil, *Curculio caryae* (Horn), in Texas. PhD. Dissertation. Texas A&M University, College Station, Texas.

Harp, S. J., and H. W. Van Cleave. 1976a. Biology of the pecan weevil. *Southwestern Entomol.* **1**: 21-30.

Harp, S. J., and H. W. Van Cleave. 1976b. Biology of the subterranean life stages of the pecan weevil in two soil types. *Southwestern Entomol.* **1**: 31-34.

Harp, S. J., and H. W. Van Cleave. 1976c. Evidence of diapause in the pecan weevil. *Southwestern Entomol.* **1**: 35-37.

Harp, S. J., and H. W. Van Cleave. 1976d. New records of natural enemies of the pecan weevil. *Southwestern Entomol.* **1**: 38-39.

Harris, M. K. 1973. Pecan weevil larval response to some temperatures while in the nut. *Texas Agric. Exp. Sta.* PR-3176. 3 pp.

Harris, M. K. 1976a. Pecan weevil adult emergence, onset of oviposition and larval emergence from the nut as affected by the phenology of the pecan. *J. Econ. Entomol.* **69**: 167-170.

Harris, M. K. 1976b. Pecan weevil infestations of pecans of various sizes and infestations. *Environ. Entomol.* **5**: 248-250.

Harris, M. K. 1978. Annual report of cooperative regional project S-88. "Bionomics and Control of the Pecan Weevil." 3 pp.

Harris, M. K. 1979. Pecan weevil distribution on pecan across the pecan belt. *Southern Cooperative Series Bull.* (In Press).

Harris, M. K., and L. Aguirre. 1978. Pecan, pecan weevil and hickory shuckworm management, 1977. *Insecticide and Acaricide Tests.* **3**: 59-60.

Hedger, G. H., R. D. Eikenbary, and R. D. Morrison. 1978. Computer prediction and control of the pecan weevil. *Pecan South* **5(6)**: 198-208.

Hinrichs, H.A. 1948. Pecan scab and pecan weevil control. *Proc. Okla. Pecan Growers Assoc.* **19**: 41-48.

Hinrichs, H. A., and H. J. Thomson. 1955. Insecticide tests for pecan weevil control. *Okla. Agric. Exp. Sta. Bull.* 450. 12 pp.

Hoyt, S. C., and E. C. Burts. 1974. Integrated control of fruit pests. *Ann. Rev. Entomol.* **19**: 231-252.

Johnson, R. H. 1978. California here they come. *Pecan South* **5(3)**: 116-117.

Kirkpatrick, R. L., W. W. Neel, and N. V. Mody. 1976. High temperature as a method of controlling pecan weevils. *J. Ga. Entomol. Soc.* **11(4)**: 293-296.

Kumpe, O., and D. Isely. 1936. Notes on biologies of nut infesting weevils. *J. Kan. Entomol. Soc.* **9**: 13-16.

Leesch, J. G., and H. B. Gillenwater. 1976. Fumigation of pecans with methyl bromide and phosphine to control the pecan weevil. *J. Econ. Entomol.* **69**: 241-244.

Leiby, R. W. 1925. Insect enemies of the pecan. *N.C. Dept. Agr. Bull.* 67 pp.

McWhorter, G. M., J. G. Thomas, M. K. Harris, and H. W. Van Cleave. 1976. Pecan insects of Texas. *Texas Agric. Ext. Serv.* MP-1270. 17 pp.

Mody, N. V., P. A. Hedin, and W. W. Neel. 1976. Volatile components of pecan leaves and nuts, *Carya illinoensis* Koch. *J. Agric. Food Chem.* **24**: 175-176.

Mody, N. V., D. H. Miles, W. W. Neel, P. A. Hedin, A. C. Thompson, and R. C. Gueldner. 1973. Pecan weevil sex attractant: Bioassay and chemical studies. *J. Insect Physiol.* **19**: 2063-2071.

Moznette, G. F. 1947. DDT for control of the pecan weevil. *Proc. Southeastern Pecan Growers Assoc.* **40**: 49-53.

Moznette, G. F., T. L. Bissell, and H. S. Adair. 1931. Insects of the pecan and how to combat them. *USDA Farmers Bull.* 1654. 59 pp.

Moznette, G. F., C. B. Nickels, W. C. Pierce, T. L. Bissell, J. B. Demaree, J. R. Cole, H. E. Parson, and J. R. Large. 1940. Insects and diseases of the pecan and their control. *USDA Farmers Bull.* 1829. 70 pp.

Neel, W. W., and M. Shepard. 1976. Sampling adult pecan weevils. *Southern Cooperative Series. Bull.* 208. 17 pp.

Neel, W. W., and P. P. Sikorowski. 1972. Pecan weevil susceptibility to fungus. *Pecan Quarterly* **6**: 14-15.

Neel, W. W., C. H. Graves, and R. E. Coats. 1976. Results of pecan weevil and spittlebug control tests in 1975. *Proc. Southeastern Pecan Growers Assoc.* **69**: 109-116.

Neel, W. W., H. R. Sterling, N. V. Mody, and L. B. Gray. 1975. Pecan weevil: Late seasonal emergence in Arkansas. *Pecan Quarterly* **9(2)**: 20-21.

Nickels, C. B. 1950. Experiments in control of the pecan weevil. *J. Econ. Entomol.* **43**: 552-554.

Nickels, C. B. 1952. Control of pecan weevil in Texas. *J. Econ. Entomol.* **45**: 1099-1100.

Nickels, C. B., and W. C. Pierce. 1947. Effect of flooding on larvae of the pecan weevil in the ground. *J. Econ. Entomol.* **40**: 921.

Osburn, M. R. 1952. Experiments for pecan weevil control in 1951. *Proc. Southeastern Pecan Growers Assoc.* **45**: 105-109.

Osburn, M. R., and W. L. Tedders. 1966. Control of the hickory shuckworm and the pecan weevil. *Proc. Southeastern Pecan Growers Assoc.* **59**: 96-100.

Osburn, M. R., V. R. Calcote, and W. L. Tedders. 1964. Low volume sprays for pecan weevil control. *Proc. Southeastern Pecan Growers Assoc.* **54**: 59-60.

Osburn, M. R., W. C. Pierce, A. M. Phillips, J. R. Cole, and G. L. Barnes. 1963. Controlling insects and diseases of the pecan. *USDA Agr. Handb.* 240. 52 pp.

Osburn, M. R., W. C. Pierce, A. M. Phillips, J. R. Cole, and G. E. KenKnight. 1966. Controlling insects and diseases of the pecan. *USDA Agr. Handb.* 240. 55 pp.

Payne, J. A., and J. M. Wells. 1974. Postharvest control of the pecan weevil in inshell pecans. *J. Econ. Entomol.* **67**: 789-790.

Payne, J. A., E. D. Harris, and H. Loman. 1976. Foliar insecticides for control of pecan weevil and hickory shuckworm. *J. Ga. Entomol. Soc.* **11**: 306-308.

Payne, J. A., R. M. Barry, E. D. Harris, S. G. Polles, and E. J. Wehunt. 1975. Pecan weevil: Field evaluation of foliar and soil pesticides. *Pecan South* **2(5)**: 136-137.

Pierce, W. C. 1963. Aerial applications of insecticides for control of pecan insect pests. *Proc. Southeastern Pecan Growers Assoc.* **56**: 60-69.

Polles, S. G. 1976. Pecan insect control studies in Georgia. *Pecan South* **3(4)**: 404-406.

Polles, S. G., and J. A. Payne. 1973. Pecan weevil: Toxicity of insecticides in laboratory tests. *J. Econ. Entomol.* **66**: 497-498.

Polles, S. G., and J. A. Payne. 1974. Pecan weevil, hickory shuckworm, and yellow pecan aphids. *Proc. Southeastern Pecan Growers Assoc.* **67**: 79-94.

Polles, S. G., J. A. Payne, and R. L. Jones. 1977. Attraction of the pecan weevil to its natural pheromone and grandlure. *Pecan South* **4(1)**: 26-28.

Polles, S. G., J. A. Payne, and E. J. Wehunt. 1973. Pecan weevil: Control with soil-applied insecticides and nematicides. *J. Econ. Entomol.* **66**: 501-503.

Pollet, D. K., and J. B. Aitken. 1978. Ground application of Sevin. *Pecan South* **5(1)**: 164-165.

Price, W. S. 1939. Observations on the pecan weevil. *Proc. Texas Pecan Growers Assoc.* **19**: 30-31.

Raney, H. G., and R.D. Eikenbary. 1968. Investigations on flight habits of the pecan weevil, *Curculio caryae* (Coleoptera: Curculionidae). *Can. Entomol.* **100**: 1091-1095.

Raney, H. G., and R. D. Eikenbary. 1969. A simplified trap for collecting adult pecan weevils. *J. Econ. Entomol.* **62**: 722-723.

Raney, H. G., and R. D. Eikenbary, and N. W. Flora. 1969. Investigations on dispersal of the pecan weevil. *J. Econ. Entomol.* **62**: 1239-1240.

Raney, H. G., R. D. Eikenbary, and N. W. Flora. 1970. Population density of the pecan weevil under Stuart pecan trees. *J. Econ. Entomol.* **63**: 697-700.

Roelofs, W. L., E. H. Glass, J. Tette, and A. Comeau. 1970. Sex pheromone trapping for red-banded leafroller control: Theoretical and actual. *J. Econ. Entomol.* **63**: 1162-1167.

Smith, R. F., and H. T. Reynolds. 1966. Principles, definitions and scope of integrated pest control. FAO Symposium. Rome, Italy. 7 pp.

Sri-Arunotai, S., P. P. Sikorowski, and W. W. Neel. 1975. Study of the pathogens of the pecan weevil larvae. *Environ. Entomol.* **4**: 790-792.

Sterling, W. L., S. G. Wellso, P. L. Adkisson, and H. W. Dorough. 1965. A cottonseed-meal diet for laboratory cultures of the boll weevil. *J. Econ. Entomol.* **58**: 867-869.

Swingle, H. S. 1931. Life history of the pecan weevil. *Ala. Exp. Sta. Ann. Rept.* **41**: 49.

Swingle, H. S. 1934. The pecan weevil. *Ala. Agr. Exp. Sta. Leaflet* 4. 4 pp.

Swingle, H. S. 1935. Control of the pecan weevil. *J. Econ. Entomol.* **28**: 794-795.

Swingle, H. S., and J. L. Seal. 1931. Some fungous and bacterial diseases of pecan weevil larvae. *J. Econ. Entomol.* **24**: 917.

Tedders, W. L. 1974. Bands detect weevils. *Pecan Quarterly* **8(3)**: 24-25.

Tedders, W. L. 1976. Pesticides for pecan pests. *Pecan South* **3(3)**: 376-379.

Tedders, W. L., and M. R. Osburn. 1971. Emergence and control of the pecan weevil. *J. Econ. Entomol.* **64**: 743-744.

Tedders, W. L., D. J. Weaver, and E. J. Wehunt. 1973. Pecan weevil: Suppression of larvae with the fungi *Metarrhizium anisopliae* and *Beauveria bassiana* and the nematode *Neoaplectana dutkyi. J. Econ. Entomol.* **66**: 723-725.

Van Cleave, H. W., and S. J. Harp. 1971. The pecan weevil: Present status and future prospects. *Proc. Southeastern Pecan Growers Assoc.* **64**: 99-111.

Van Cleave, H. W., D. J. Boethel, S. J. Harp, and A. L. Anderson. 1969. Chemical control field studies on the hickory shuckworm and pecan weevil in Texas. *Proc. Texas Pecan Growers Assoc.* **48**: 45-47.

EXTENSION APPROACHES TO PECAN PEST MANAGEMENT IN ALABAMA AND GEORGIA

John R. McVay
Alabama Cooperative
Extension Service
Auburn University
Auburn, Alabama

H. C. Ellis
Georgia Cooperative
Extension Service
University of Georgia
Tifton, Georgia

INTRODUCTION

In Alabama and Georgia, as with most other States, the active organization and implementation of crop pest management systems has been carried out by the respective cooperative extension services. Relying upon base data collected and provided by their research counterparts, extension scientists have been charged with the responsibility of assimilating this information into a complete program that can be put to practical use by the grower. This often includes such diverse problems as refinement of survey techniques, definition of economic or action threshold levels, preparation and subsequent evolution of survey and reporting forms, training of survey personnel, and the intensive education of growers in the use of the various pest management tools.

Thus extension scientists in Alabama and Georgia have developed sound pest management programs dealing with a variety of crops. In Alabama, these include cotton, peanuts, soybeans and pecans with work progressing on a tomato program. Georgia has active programs dealing with cotton, tobacco, peanuts, soybeans and pecans. These programs require close cooperation among workers within the various plant protection disciplines; the county, district and state staffs within the cooperative extension service; and considerable direct input from the various research staffs.

Pecan pest management programs are of recent origin in both states covered in this chapter. Active field implementation of the programs began in 1977. Both programs are geared for full state-wide implementation in 1979, as growers become aware of the advantages offered them by such programs.

Due to differing initial approaches in the two states, the remainder of this chapter will treat each program separately and possibly offer some contrast and clarification to others who are beginning to develop similar programs.

THE ALABAMA PECAN PEST MANAGEMENT PROGRAM

In 1977, the Alabama Cooperative Extension Service and Auburn University initiated the first year of field work on a Pilot Pecan Pest Management Program. This marked the beginning of a three-year effort to develop and implement various interdisciplinary pest management techniques leading to the establishment of a statewide, self-supporting pecan pest management program. For the study, a three-year grant was provided by USDA-Extension Service.

A program was planned which would take into account all recognized pecan pests, including insects, disease, weeds and vertebrate animals. Following is an account of how the program was organized and put into effect as well as future plans. In keeping with accompanying chapters, only insect and disease management programs are described in detail.

Organization

Field operations are conducted by, or under the supervision of the pest management specialist-pecans. Others involved with direct field input are the extension entomologist-horticulture, extension plant pathologist-nematologist, and extension weed scientist. Economic analysis is handled by the extension economist-pest management in cooperation with those already mentioned. Several steering committees (one for each discipline involved) and an advisory committee serve as reservoirs of valuable information gleaned from first-hand knowledge of pecan production and pests. These committees are composed of personnel from the extension service specialist and county staffs, Auburn's research staff, USDA research, USDA-APHIS, and the Alabama State Departments of Agriculture and Conservation, as well as pecan growers from the project area. Meeting once yearly, these committees review the problems and progress of the past year and make decisions adjusting the specific conduct of the program for the coming year.

In 1977 the program was limited to Baldwin and Mobile counties in southwestern Alabama. Due to the nature of the program, in its infancy, it was decided that pest management (PM) on a demonstrational basis would be best suited for the development of techniques, infestation thresholds (action levels) and other pertinent aspects. A total of twelve growers, six in each county, were asked to place a portion of their orchard(s) under the PM program and to continue to conduct the remainder of their orchard operations as they normally would. This non-PM area would serve as a check or control area to compare PM practices with a normal operation. The PM orchards ranged in size from 5 to 17 acres but usually encompassed the area that the cooperator normally treated with a single 500 gallon tank of spray mixture. The 1977 season was spent working with this limited number of growers and evaluating different methods that have shown promise in research. A great deal of the work early in the program has been devoted to the evaluation of monitoring techniques and the establishment of guidelines for economic injury or treatment levels of pest infestations since these treatment levels were generally lacking for most pecan pests. In developing economic pest population injury levels, a number of monitoring and control techniques have been evaluated including blacklight trap catches, egg counts, honeydew ratings, beneficial complex evaluations, selective pesticides, blacklight suppression techniques, attractants, traps, season and climatic population peaks, leaf and twig growth and development stages, late season foliage retention dates, nut quality and yield, mechanical barriers, repellents, and hunting and trapping of birds and animals. Those methods and techniques that are both economically and operationally feasible at the grower level were to be adopted in the 1978 growing season and made available to all growers in the two-county project area.

Insect Pest Management

The insect pest management phase of the program in 1977 was designed to show, through result demonstrations, that an integrated program can replace the intensive, preventative schedule of chemical treatments currently in widespread use. The alternative program made effective use of scouting, threshold or action levels, blacklight traps, beneficial organisms, selected chemical treatments, and other factors which enabled participating growers to realize financial savings without reducing the degree of control over the insect pests in question.

Target pest species included the pecan nut casebearer, *Acrobasis nuxvorella* Neunzig; a yellow aphid, *Monelliopsis nigropunctata* (Fitch); the hickory shuckworm, *Laspeyresia caryana* (Fitch); the black pecan aphid, *Tinocallis caryaefoliae* (Davis); and the black-margined aphid, *Monellia caryella* (Fitch). Observations were made on the presence of other pest species as well as that of beneficial arthropods. Results, as presented, are broken down by orchard and/or pest except where demonstrations were designed to compare methods or control agents.

The basis of most pest management practices was twice-weekly scouting which, with pre-set, but flexible action levels allowed the grower sufficient notice as to the times when treatment was needed. Specialized alternatives consisting of chemical treatments and population suppression techniques were utilized in certain PM orchards.

Pecan Nut Casebearer. All twelve PM orchards were scouted twice-weekly during the time of peak nut casebearer activity. During this period (late April-late May), a visual observation of 20 nut clusters and terminals was made on each scout tree for eggs, larvae and/or damage. Counts were broken into percentages for use with the action level established at 6% infestation.

The pecan nut casebearer population was not found to be economically damaging in any PM orchard during the 1977 growing season. Of the twelve orchards, minor infestations were discovered in three and none in the remainder. Infestations never exceeded 0.63% in one orchard or 1.25% in the other two. As the action level had been pre-set at 6% infestation and an extremely heavy nut set was evident in all PM orchards, no control measures were suggested or taken.

Hickory Shuckworm. Two approaches were taken in the management of the hickory shuckworm. The first consisted of blacklight trap surveys and visual observation of damage. The second approach was an attempt to suppress the population below economic injury levels by the use of blacklight traps (Tedders *et al.* 1972) suspended in the tree canopies at the rate of two traps per acre. These orchards were also scouted for the presence of other insect pests and beneficials.

The light traps in one of the eight PM orchards in which the scouting and survey approach was utilized were stolen early in the season and no useable data was obtained. In the other seven orchards, scouting trips were made twice-weekly and the survey traps emptied each time. The trap contents were screened for hickory shuckworm adults and captures were recorded as moths per trap per night. Based on this data, the action level of nine (9) moths per trap per night was never reached in any of the PM orchards. (Fig. 1 and 2).

A significant infestation did develop in most orchards in the latter part of September. This was not discovered until samples were taken during the first week in October from both PM orchards and control orchards. In this case, control orchards were those in which each grower followed his normal pesticide spray schedule. Sample size was 25 nuts with shucks per tree from six trees in each orchard for a total of 150 nuts/orchard. These were mixed and the first 100 taken at random from the holding container were inspected for signs of shuckworm infestation. Results are presented in Fig. 3. Average infestation in PM orchards was 39% and in control orchards, 37%. This late season influx was not detected due to removal of the survey traps in the third week of September. Traps were removed then because of early nut maturity and shuck split, after which time, no insecticides can be legally applied. This late infestation did not do significant damage to the pecan crop, again due to the early maturation of the nut crop.

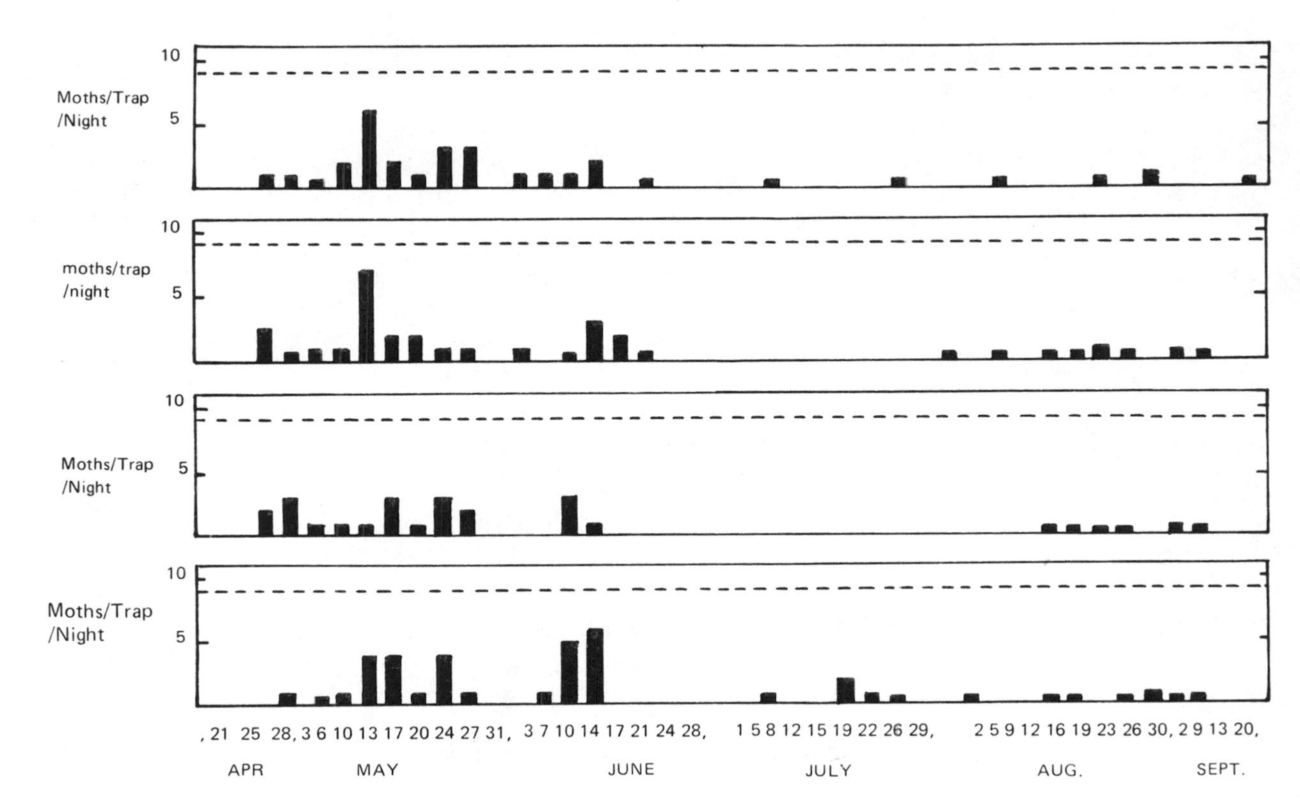

Figure 1. The number of adult hickory shuckworms captured in PM orchards 01M, 02M, 03M, and 04M, respectively, throughout the 1977 growing season. The broken line represents the action level of 9 moths/trap/night. [Adapted from: McVay *et al.* (1977)].

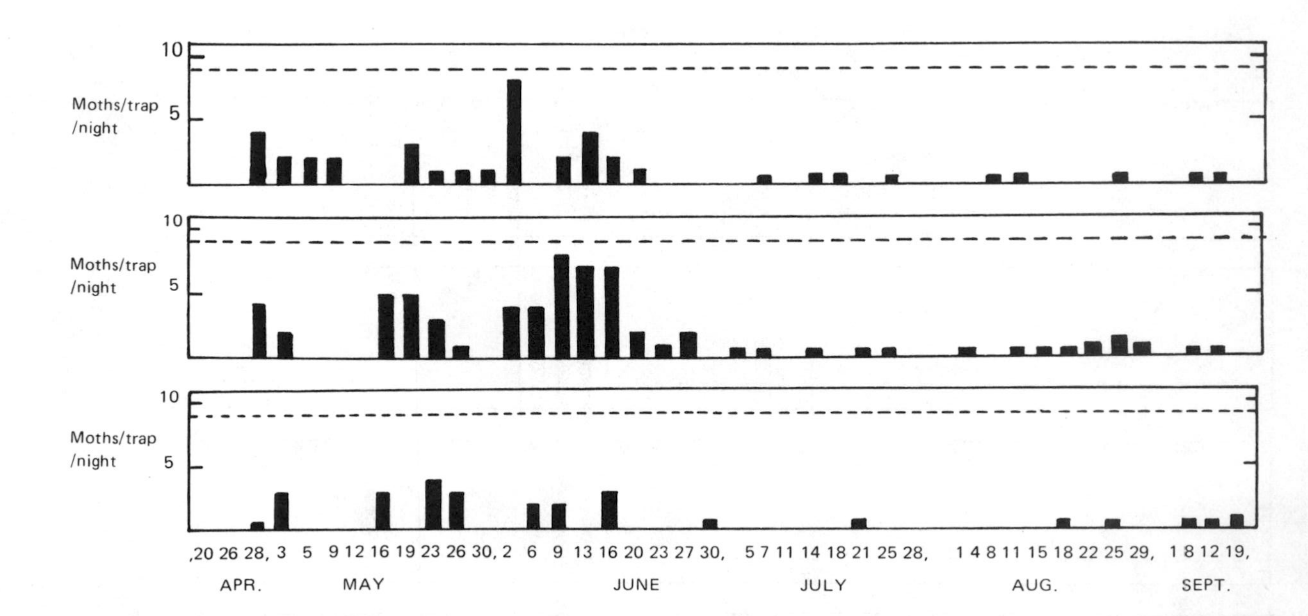

Figure 2. The number of adult hickory shuckworms captured in PM orchards 01B, 02B, and 04B, respectively, throughout the 1977 growing season. The broken line represents the action level of 9 moths/trap/night. [Adapted from: McVay *et al.* (1977)].

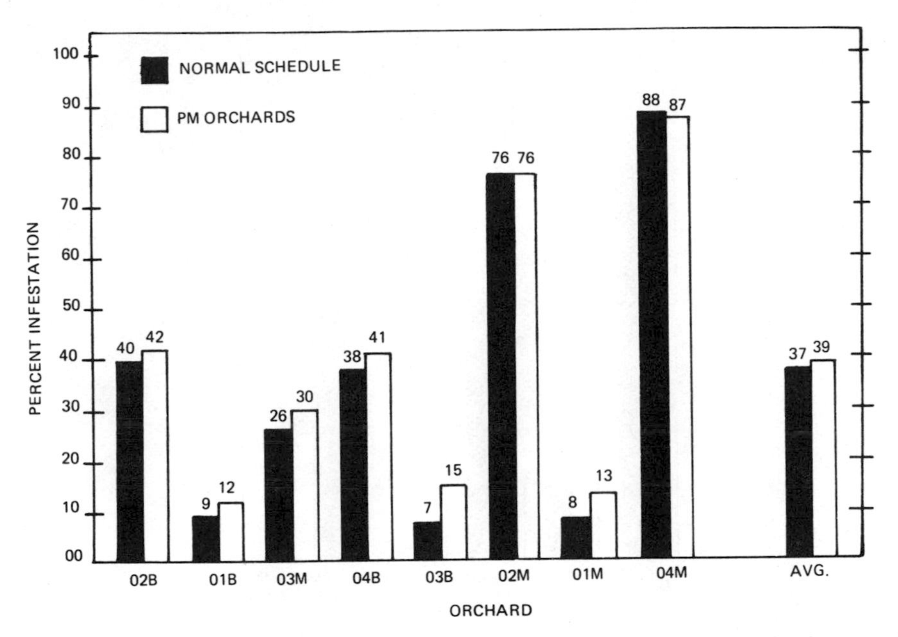

Figure 3. Percent infestation of shucks by the hickory shuckworm in orchards scouted under the pest management system as compared to orchards held under a normal insecticide spray program. [Taken from: McVay *et al.* (1977)].

Four PM orchards were involved in the second approach in which blacklight traps were utilized to suppress the shuckworm population. These traps were allowed to remain in the orchards until the first week in October, and as a result, reduced the infestation dramatically. Average infestation in three of the PM orchards was 26% as compared to 52% in control orchards (Fig. 4). One of the four was not included in the final data due to no infestation in either the PM or control orchard. This reduction is doubly significant in that literature reports that an infestation of less than 30% will not be economically damaging (Tedders *et al.* 1972).

Yellow Aphids. As indicated earlier, there are two species of yellow aphids which attack pecans. The smaller of the two, *M. nigropunctata,* occurs predominantly in the spring and early summer. The larger, black-margined aphid, *M. caryella,* is more damaging to the tree and deposits considerably more honeydew which supports the growth of sooty mold fungus. It has been commonly accepted that this aphid occurs primarily in the late summer, following a lapse in July when virtually no aphids are present. This was largely true of the 1977 growing season with the exception that a well mixed population of both species became obvious in early June requiring alteration of the pre-set action level.

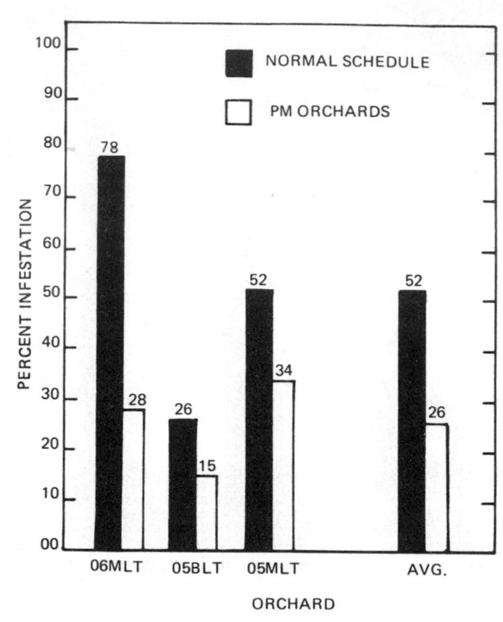

Figure 4. Percent infestation of pecan shucks by the hickory shuckworm in PM orchards equipped with blacklight traps at 2/acre as compared to orchards held under normal insecticide spray program. [Taken from: McVay *et al.* (1977)].

Action levels were previously set at 30 aphids/compound leaf during the spring and early summer when the smaller *M. nigropunctata* was the predominant species. For the late summer bloom of black-margined aphids, it was set at 10/compound leaf. Both action levels worked very well in the so-called normal course of events. However, when scouting surveys revealed that a well mixed population (est. 50/50) was found in all orchards in early June, the action level of 30 was lowered to 15 aphids/compound leaf. This level worked extremely well as controls seem to have been applied at the correct time to keep the populations under control (Fig. 5 and 6). All growers applied phosalone as the insecticide of choice when necessary.

Three PM orchards equipped with light traps for hickory shuckworm suppression were treated with soil-applied systemic insecticides. These materials, aldicarb and disulfoton, were applied in orchards utilizing trap suppression of the shuckworm because they are specific materials and the commonly used phosalone would affect both aphid and shuckworm, thereby negating the usefulness of the blacklight traps. Aldicarb was applied at the rate of 0.45 lb. ai/tree and disulfoton at 1.12 lb. ai/tree. Both are 15% granular products and were applied in 6 ft. bands on two sides of the tree using a grain drill.

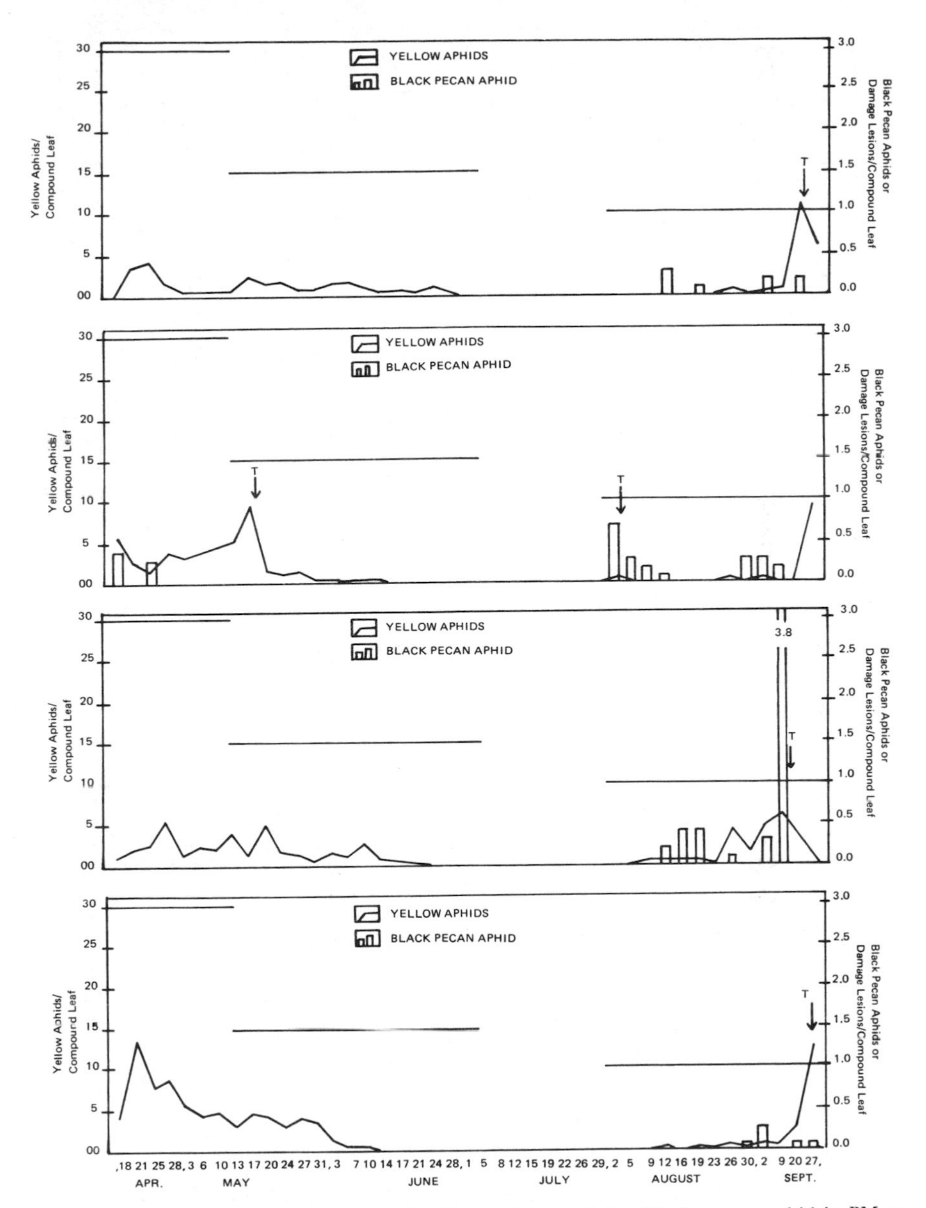

Figure 5. The infestation levels of yellow aphids and the black pecan aphid in PM orchards 01M, 02M, 03M, and 04M, respectively, throughout the 1977 growing season. The horizontal lines represent action levels (30, 15, and 10 for yellow aphids and 1.0 for the black pecan aphid). T represents treatment dates. [Adapted from: McVay *et al.* (1977)].

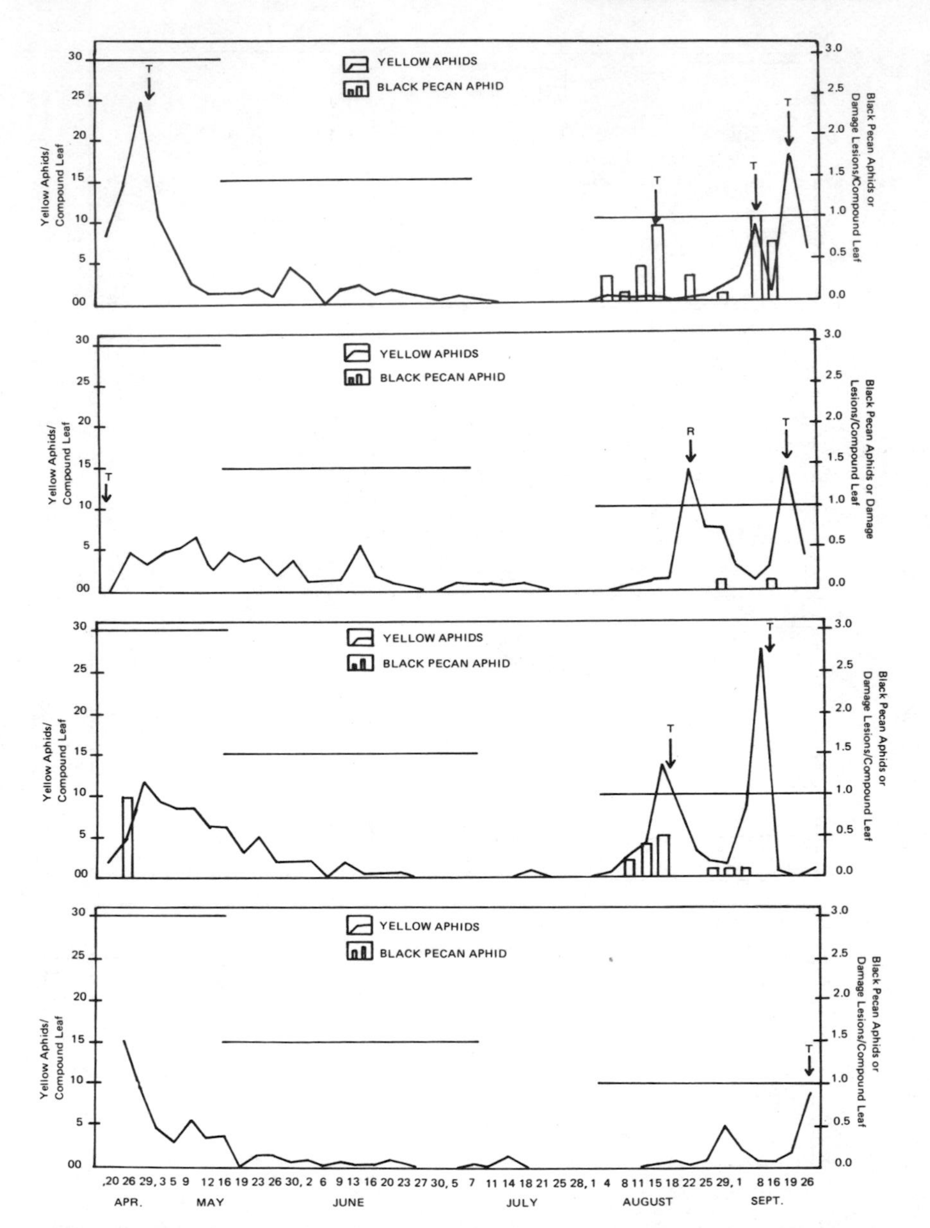

Figure 6. The infestation levels of yellow aphids and the black pecan aphid in PM orchards 01B, 02B, 03B, and 04B, respectively, throughout the 1977 growing season. The horizontal lines represent action levels (30, 15, and 10 for yellow aphids and 1.0 for the black pecan aphid). T represents treatment dates; R represents day of heavy localized rainfall. [Adapted from: McVay *et al.* (1977)].

Excellent to complete control of all aphids was obtained in all three orchards (aldicarb in two, disulfoton in one) following application in early July (Fig. 7). Application was by grain drill and sufficient rainfall followed to allow the materials to enter the root systems of the trees very quickly.

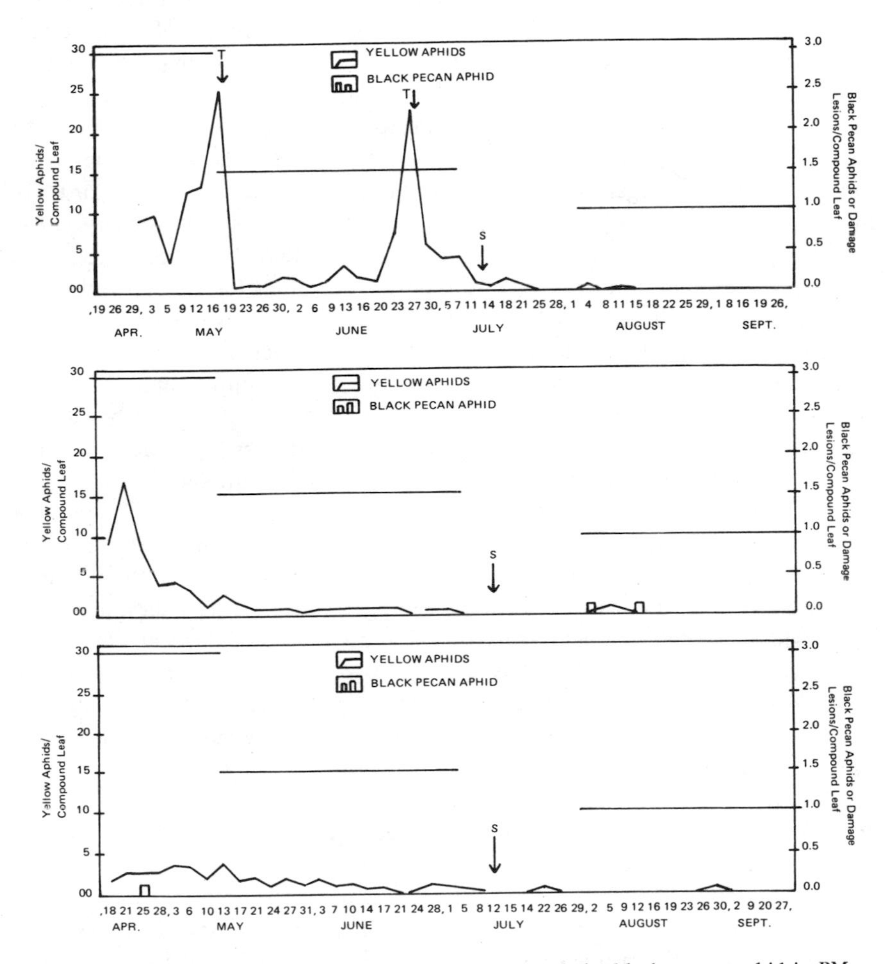

Figure 7. The infestation levels of yellow aphids and the black pecan aphid in PM orchards 06BLT, 05MLT, and 06MLT, respectively, throughout the 1977 growing season. The horizontal lines represent action levels (30, 15, and 10 for yellow aphids and 1.0 for the black pecan aphid). T represents treatment dates; S represents date of systemic application. [Adapted from: McVay *et al.* (1977)].

Black Pecan Aphid. The black pecan aphid is a late summer pest that can cause serious, premature defoliation of the tree. In 1977, scouting showed this pest present throughout most of the growing season but not in harmful numbers until the late season period beginning around the first of August. Although an easily controlled pest, the black pecan aphid is so serious a threat as a defoliator that an action level of one (1) aphid, damage lesion, or combination of the two per compound leaf was deemed necessary for the months of August and September. This proved to be the correct approach as it was found that the pest population could quickly get out of hand in an orchard situation.

Scouting proved effective in the management of this insect as it was with the yellow aphids. The systemic insecticide treatments discussed earlier were completely effective against this pest also. Population levels in PM orchards are given in Fig. 5 - 7.

Other Pests. Several pest species other than the ones already discussed were encountered during the field surveys during 1977. Among these were stink bugs, leaffooted bugs, leafminers, fall webworms, spittlebugs, phylloxera, and mites. Probably the most damaging are the pecan leaf scorch mite, *Eotetranychus hicoriae* (McGregor), the leaffooted bug/stink bug complex, and the pecan serpentine leafminer, *Nepticula juglandifoliella* Clemens. All should be fit into the scheme of the pest management system as it continues to develop. Two PM orchards were treated once for infestations of pecan leaf scorch mite. Sulfur was chosen as the control agent due to its toxicity to the mite and lack of impact on the beneficial arthropod complex on pecans. The senior author believes that the leaffooted bug/stink bug complex caused a large part of an August nut drop which was attributed by many to physiological factors. This has not been confirmed but population and damage surveys are under development for future use. The serpentine leafminer has become a very serious pest where heavy pesticide usage has been practiced. Insecticides presently registered for use on pecans have virtually no effect on this insect but there are parasites that keep it under control where insecticide pressure is not heavy (Tedders 1978b).

Beneficials. The complex of beneficial arthropods to be found in an orchard situation is very large and of extremely varied composition (Tedders 1976 and 1978a). Most of the beneficials observed in 1977 fed upon the aphid complex attacking pecans but each pest does have its own natural enemies which aid the grower in a pest management system. Notable beneficials observed were several species of lady beetles, three species of lacewings, syrphid flies, damsel bugs, and several spiders (Table 1). Altogether, this complex does a good job of suppressing several of the pest species present in pecans.

Disease Pest Management. Pecan scab *(Fusicladium effusum)* is the principal disease threat to pecan production consequently, the major emphasis was given to scab control. Twelve result demonstrations were initiated in the two-county

Table 1.
Common and scientific names of some
of the more important predators attacking pecan pests.

Common Name	Scientific Name	Order and (Family)
ashgrey lady beetle[a]	*Olla abdominalis* (Say)	Coleoptera (Coccinellidae)
convergent lady beetle	*Hippodamia convergens* (Guerin-Meneville)	Coleoptera (Coccinellidae)
twicestabbed lady beetle	*Chilocorus stigma* (Say)	Coleoptera (Coccinellidae)
an assassin bug	*Sinea spinipes* (Herich-Schaeffer)	Hemiptera (Reduviidae)
an assassin bug	*Zelus exsanguis* (Stal)	Hemiptera (Reduviidae)
spined soldier bug	*Podisus maculiventris* (Say)	Hemiptera (Pentatomidae)
goldeneye lacewing	*Chrysopa oculata* (Say)	Neuroptera (Chrysopidae)
a lacewing	*Chrysopa rufilabris* Burm.	Neuroptera (Chrysopidae)
a lacewing	*Allochrysta virginica* (Fitch)	Neuroptera (Chrysopidae)
a syrphid fly	*Allograpta obligua* (Say)	Diptera (Syrphidae)

[a]Not an approved common name.

area which compared with the presently recommended commercial spray program (8-10 applications) to an abbreviated (5 sprays) program. Also tested were different rates of fungicides and the use of a combination of the two most widely used materials. An explanation of the treatments and a comparison of the spray schedules utilized are shown in Tables 2 and 3, respectively. Applications were made as timely as possible. Thorough coverage and correct rates were achieved in all cases. Orchard areas in which the grower continued to practice his normal operation (commercial program) were utilized as control or comparison orchards. Factors such as air drainage, physical configuration, and the surrounding vegetation were taken into consideration and suggestions made to the grower for improvement of air movement in and around orchards.

Ratings were made of scab incidence on both foliage and nuts (shucks) at intervals throughout the growing season. Infection of foliage was rated in June, July, and August, while shuck or nut infection was determined in July, August, and September. Ten percent of the trees in each orchard were sampled and five compound leaves were taken at random from each tree. Nut samples were taken randomly in a circular pattern around the tree. Leaf samples were taken in the same manner with the leaves taken from the middle area of the current year's terminal growth. Ratings were made of each sample on a numerical scale of 0-4

Table 2.
Treatments applied to pest management demonstration orchards in 1977 for control of pecan scab, *(Fusicladium effusum).*[a]

NUMBER	SCHEDULE	FUNGICIDE	RATE[b]
IA	Abbreviated	Du-Ter	2X
IB	Abbreviated	Du-Ter + Benlate	1X 1X
IIA	Abbreviated	Du-Ter	2X (1st) + 1X (remainder)
IIB	Abbreviated	Du-Ter	2X (1st) + 1X (remainder)
IIIA	Abbreviated	Du-Ter	1X
IIIB	Commercial	Du-Ter	1X

[a]Taken from: McVay *et al.* (1977).

[b]1X rates are based on the 1977 Commercial Pecan Spray Schedule which recommends 2 lb. of Du-Ter/100 gal. of spray mixture, applying 12-18 gal. to each tree (approx. 33 trees per 500 gal. tank). Benlate 1X rate = 1/2 lb./100 gal.

where 0 equals no infection; 1 equals trace to 10%; 2 equals 11 to 25%; 3 equals 26 to 50%; 4 equals 51 to 100%. The resulting data were broken into percentage of samples in each category and a disease severity index (D.S.I.) calculated with the following formula:

$$\text{D.S.I.} = \frac{(n)\,0 + (n)\,1 + (n)\,2 + (n)\,3 + (n)\,4}{(nt)\,4}$$

Where: n = number of samples in a grade
nt = total number of samples

Both the commercial spray program and the PM abbreviated program provided adequate control of pecan scab during the 1977 growing season. Results are presented in Table 4. Analysis of variance revealed that results were inconclusive when comparisons of the various PM treatments were made with the respective control orchards. In all but three cases there were no significant differences between the two rates of infection. In one orchard (Trt. IA) the abbreviated PM schedule provided significantly better control of scab than did the commercial program. The reverse was true of a second, similar test (Trt. IIIA). Note that the treatments were identical except for the fungicide rates used (Table 2). The third test in which there was a

Table 3.
The recommended commercial pecan spray schedule
and the abbreviated pest management schedule in 1977.[a]

COMMERCIAL SCHEDULE		ABBREVIATED SCHEDULE	
Application Name or Number	Approximate Date	Application Name or Number	Approximate Date
1. 1st Prepollination (1)	April 1	1. Bud-Break (1)	March 23
2. 2nd Prepollination (2)	April 15	2. Prepollination (2)	April 13
3. Pollination (3)	April 29	3. Pollination (3)	May 4
4. 1st Cover (4)	May 13		
5. 2nd Cover (5)	June 3		
6. 3rd Cover (6)	June 24	4. Nut Expansion I	July 1
7. 4th Cover (7)	July 15	5. Nut Expansion II	July 22
8. 5th Cover (8)	Aug. 5		
9. 6th Cover (9)	Aug. 26		

[a]Taken from: McVay *et al.* (1977).

difference concerned one of the orchards in which Benlate (benomyl) and Du-Ter (triphenyltin hydroxide) were applied as a tank mix on an abbreviated schedule (Trt. IB). In this case, the commercial spray program with Du-Ter alone as the chemical agent produced significantly better control than did the tank mix.

Results were inconclusive in that neither spray program consistently provided better scab control. It was noted that a more than desirable rate of infection was experienced with both programs in most cases; however, no commercial loss was evident.

Weed Pest Management. This program involves using herbicides to obtain a 10 to 12 foot weed free band within the tree row and also the use of herbicides to chemically frost mowed vegetation between the rows just prior to nut fill. The weed free band within the row eliminates weed competition, increases harvesting efficiency, prevents mechanical mower damage to the tree trunks, and eliminates disking which destroys feeder roots and spreads crown gall *(Agrobacterium tumefaciens).* Chemical frosting of mowed vegetation between the rows reduces water competition during nut fill.

Major weed problems in pecan orchards within these two counties are: bahiagrass, bermudagrass, briars, prickly sida, curly dock, cheat, rescuegrass, chickweed, little barley, morning glory, large crabgrass, and goosegrass. This spectrum of weeds includes winter annuals and perennials, as well as summer deep-rooted weeds such as bahiagrass, prickly sida, briars, curly dock,

Table 4.
Final disease severity index rating
in pest management (PM) and control orchards (CK).[a]

Treatment	County	Disease Severity Index (D.S.I.)			
		Foliage		Nuts	
IA	Mobile	PM	.33	PM	.50
		CK	.29	CK	.52
	Baldwin	PM	.28[c]	PM	.20[c]
		CK	.39	CK	.27
IB	Mobile	PM	.28[c]	PM	.27
		CK	.42	CK	.33
	Baldwin	PM	.17	PM	.30
		CK	.19	CK	.19
IIA	Mobile[b]	PM	.50	PM	.60
		CK	.56	CK	.64
	Baldwin	PM	.30	PM	.24
		CK	.25	CK	.28
IIB	Mobile	PM	.37	PM	.54
		CK	.40	CK	.61
	Baldwin	PM	.24	PM	.18
		CK	.26	CK	.19
IIIA	Mobile	PM	.23	PM	.30[c]
		CK	.28	CK	.20
	Baldwin[b]	PM	.31[c]	PM	.40
		CK	.46	CK	.44
IIIB	Mobile	PM	.32	PM	.53
		CK	.39	CK	.48
	Baldwin	PM	.49[c]	PM	.27
		CK	.27	CK	.24

[a]Taken from: McVay *et al.* (1977).

[b]Remained on commercial schedule on advice of specialist.

[c]Paired D.S.I. values are significantly different; ($P \leq 0.05$).

and goosegrass. Demonstrations during 1977 dealt with only summer annuals and perennials but will include both summer and winter weeds for the remainder of the project.

Herbicides applied within the tree row and used as a chemical frost between rows, in combination with mechanical mowing between rows, provided an excellent orchard floor for efficient mechanical harvesting. The herbicide programs for 1977 provided commercially acceptable control for annual broadleaf and grassy weeds. Control of perennial grasses, bahiagrass in particular, was slightly less than desired. Perennial broadleaf weed control was acceptable.

Vertebrate Pest Management. The principal vertebrate pests of pecans in the project area are bluejays, crows, and squirrels. Therefore, bird and animal pest management practices are aligned with two basic principles: (1) prevention of the pest involved to establish use patterns in the pecan orchard, and (2) early harvesting in order to reduce the time or number of months that the nut is available to the pest for predation.

Electronic scare devices (AV-Alarm model ST-4[1]) were installed in two orchards with a history of heavy losses from crows and bluejays. These scare devices have been an immense success. Both species have continually avoided the orchards equipped with scare devices. Flocks of crows approaching PM orchards have been observed to veer immediately when one of the devices goes off in the orchard.

Attempts at live trapping of squirrels were generally unsuccessful. Efforts to organize and promote direct reduction (primarily by hunting) of this pest appears to show some promise for squirrel control in pecan orchards.

Economic Evaluation

It was hoped that several objectives could be met by initiating pest management techniques in an orchard. An important objective was to improve the pecan grower's net profit. Pest management practices attempt to accomplish this goal by altering automatic, season long spray schedules and by taking advantage of all available technology to hold pests below economically damaging levels. Orchard monitoring by a scout was used to obtain optimum timing of selective spray applications and specific pesticides were used only on an "as needed" basis.

Costs for pecan pest management occur in the form of the scouting fee and the cost of blacklight traps and their power supply. Yearly economic benefits can be obtained by reducing the number of pesticide applications.

Pest management orchards using the abbreviated or modified-abbreviated fungicide schedules averaged 3.3 less fungicide applications. Even though the strict abbreviated schedule called for 5 applications, a decision was made to use additional fungicide applications in some of the orchards where the "budbreak" spraying was applied late, thus raising the average to 5.7.

In a conventional spray program, insecticide and fungicide are tank-mixed each spraying and applied together. Separate applications of insecticide sprays are seldom made in the two pilot counties because the conventional schedule provides adequate control of most insects. Using pest management techniques,

---

[1]AV-ALARM Corporation, P.O. Box 2488, Santa Maria, California 93454.

however, it's highly probable that the need for timely insecticide applications may not coincide with the fungicide schedule (especially the abbreviated schedule). To take advantage of pest management decisions, insecticide must sometimes be applied in a special spray. Therefore, PM orchards showed a reduction of only 2.0 total spray trips through the field.

When dollar values are given to these spray reductions, per acre savings can be calculated as shown in Table 5. Savings on insecticides accounted for over half of the approximate $50.00 per acre total savings. Scouting expense associated with the PM orchards cost about $7.00 per acre for 5 months of scouting. Initial investment in blacklight traps may run from $50.00 to $100.00 per orchard depending on the availability of a power source, but yearly operating costs for the traps are nominal. No significant yield or nut quality difference was detected between PM and control orchards during 1977, but actual yields were slightly higher in the PM orchards.

The $50.00 cost reduction is by no means guaranteed for every year. In pest management spraying is done on an "as needed" basis; so in years when pest pressures are high, the difference would not be nearly so great. In any year, however, scouting should provide a top level management tool to pecan growers who want to stay abreast of the current pest situation in their orchards.

Program Adjustments

In a December 1977 meeting of the steering and advisory committees, the results of 1977 were reviewed and plans finalized for the 1978 growing season. The most practical aspects of the program were chosen for presentation to the growers of the program area in county meetings. Slight changes were made in some threshold levels (Table 6) and a major revision was completed on the system of reporting infestation levels.

Table 5.

Dollar values associated with spray reductions in pest management demonstration orchards in 1977.[a]

Materials	Net Difference	Value/Acre	Savings/Acre
Insecticide	7.4	$ 3.50	$ 25.90
Fungicide	3.3	5.25	17.33
Spray Cost	2.0	3.00	6.00
Total Savings Per Acre			$ 49.23

[a]Taken from: McVay *et al.* (1977).

Table 6.
Treatment action levels
for the major arthropod pests of pecans in Alabama.

PEST	ACTION LEVEL	EFFECTIVE DATES
Yellow Aphids	30/compound leaf 15/compound leaf 10/compound leaf	Budbreak - June 1 June 1 - August 1 August 1 - Shuck Split
Black Pecan Aphid	1/compound leaf	August 1 - Shuck Split
Pecan Nut Casebearer	6% of nut clusters infested	Budbreak - June 15
Hickory Shuckworm	8 adults/trap/night	June 15 - Shuck Split
Pecan Weevil	40/tree	July 15 - Harvest
Pecan Leaf Scorch Mite	Moderate-Heavy Population	Season Long

As a result of the grower meetings, the beginning of the 1978 growing season found a total of 25 growers with ca. 2000 acres of pecans taking advantage of the PM program. There are presently four scouts working in the two-county area, receiving $7 per acre for their services. These scouts were trained at a "Pecan Scouting Short Course" held in Fairhope, Alabama. Here, they learned the basics of survey techniques, pecan growth and development, and recognition of pest and beneficial organisms found in pecan orchards.

A notable addition to the 1978 program is the evaluation of survey techniques and threshold levels concerning the pecan weevil, *Curculio caryae* (Horn). Two demonstrations have been initiated in Monroe County, Alabama in which prediction equations and trapping methods (Boethel *et al.* 1976a, b; Eikenbary *et al.* 1977 and 1978) are to be evaluated for applicability in Alabama orchards. This insect will, as a matter of course, have to be included in the statewide program to be initiated in 1979.

Conclusions

The 1977 growing season proved to be a good year for the Alabama Pecan Pest Management Project. In all discipline areas several things were accomplished, several new concepts proven and some weak points discovered and corrected.

Insect pest threshold or action levels proved generally effective and reasonable with the possible exception of the hickory shuckworm. This level was reviewed and revisions made for the coming year (Table 6). In the same context, scouts are being required to empty survey traps three times weekly in an attempt to better sample the pest populations present. The blacklight trap suppression techniques for the hickory shuckworm proved extremely effective, as did the systemic insecticides applied for aphid control. Other, foliar applied, systemic materials are being evaluated in 1978 to complement these practices. One of the greatest needs for pecan pest management is the systematic identification and cataloging of all beneficial arthropods found in pecan orchards.

Fungicide programs proved effective in 1977 and demonstrations have continued through the 1978 growing season which utilize most of the treatments from 1977 as well as some additional treatments. Data accumulation and ratings have been improved in an effort to pinpoint any possible decline in quality or yield that might be due to differences in scab infection under each program.

Vertebrate pest management was somewhat successful as related to crows and bluejays. The 1978 season will see attempts to remove squirrel populations prior to nut maturity and additional repellant (both chemical and mechanical) procedures used to manage bird pest populations.

The overall program was very successful in its first year of operation. Techniques and information will continue to undergo revision as necessary in years to come.

The program is in effect in 1978 for all interested growers in the affected two-county area of Alabama. Plans call for statewide expansion which will give all Alabama pecan growers the opportunity to utilize pest management practices in their operations. A bi-weekly newsletter, begun in 1977, will continue to provide up-to-date information throughout each growing season and plans are being made for television and radio broadcast of pest and production information. The pecan producer stands to benefit greatly from this program and the intensive program of pest management education will continue to be strengthened.

THE GEORGIA PECAN PEST MANAGEMENT PROGRAM

In Georgia, the most recent pecan pest management efforts were initiated on a demonstrational basis in 1977, and expanded to a grower-supported pilot action program in 1978. The program will be offered to growers on a broader scale in 1979 as a sub-project of the Expanded Pest Management Program of the Georgia Cooperative Extension Service. When fully operational, it will function as a grower-supported program through County Pest Management Associations. Educational programs will offer pest management information and techniques to individual growers and consultants. The interdisciplinary program considers all aspects of pecan production although major program emphasis has thus far been placed on management of insect and disease pests.

Since pesticides continue to be the primary manipulative instrument for pest management, the basic program approach has been to optimize timing and consequently the efficacy of pesticide (insecticide and fungicide) applications. Timing of applications is based on accurate, consistent, and objective assessments of pest populations (scouting). The first task was to establish and evaluate sampling procedures and treatment guidelines (action thresholds). Effectual monitoring techniques and action thresholds are mandatory for development of effective pecan scouting guidelines. Standardized survey procedures and action thresholds have not generally been available for pecan pests.

1977 Program

To provide the needed base for a pecan pest management program, tentative sampling procedures and action thresholds (for insect pests) were proposed and evaluated in pecan orchards in 1977. Demonstrations of general pest management principles were conducted by extension personnel in five counties. Specific monitoring techniques and action thresholds were evaluated in six orchards totaling 350 acres. The sampling methods utilized included general orchard surveys and trapping.

Insect Management. Bi-weekly orchard surveys were conducted to determine infestation levels of major insect pests of pecan. These included a yellow aphid, *M. nigropunctata,* the black margined aphid, the pecan nut casebearer, the black pecan aphid, mites, primarily the pecan leaf scorch mite, and the pecan spittlebug, *Clastoptera achatina* Germar. The following survey procedures were used in orchard sampling in 1977. (1) Trees in all segments of each orchard were sampled to avoid missing "hot spots" (infestations of pests such as mites and aphids are frequently localized); (2) Trees of each major variety (cultivar) were sampled because of varietal preference exhibited by some pests; (3) A minimum of 5% of the trees in each orchard were surveyed. Additional trees were surveyed as pest levels approached action thresholds; (4) A minimum of four compound leaves and four terminals (nut clusters) per tree were sampled.

Approved traps were used for early detection of the hickory shuckworm and the pecan weevil. Blacklight traps were used to monitor shuckworm populations. These are effective survey tools for hickory shuckworm and other lepidopterous pests of pecan (Tedders and Osburn 1966). Emergence of adult pecan weevils was monitored with cone-cage traps and burlap bag-band traps. These methods have been described by several researchers (Raney and Eikenbary 1969; Polles and Payne 1972 and 1973; Nash and Thomas 1972; West and Shepard 1974; Tedders 1973 and 1974). These sampling techniques proved effective but were slightly modified in 1978 to reduce scouting time.

Table 7.
Treatment guidelines utilized for arthropod pests in Georgia Pecan Pest Management Program in 1977.

INSECT	ACTION THRESHOLD
Yellow Aphids	Prior to August 1 - An average of 20 aphids/compound leaf or when honeydew is rated heavy. After August 1 - An average of 10 aphids/compound leaf or when honeydew is rated moderate.
Black Pecan Aphid	An average of 1 aphid and/or damaged area/compound leaf or when damage is observed frequently and generally throughout an orchard.
Mites	When mites are observed and scorch damage first appears on foliage of low limbs.
Pecan Nut Casebearer	When 5% of nut clusters have casebearer eggs and/or are infested by casebearer larvae.
Hickory Shuckworm	When there is an increase in number of shuckworm moths in BLT catches for 3 consecutive trapping periods and when 7 or more moths are captured in any one BLT.
Pecan Weevil	When sampling techniques (cone cage traps, trunk traps, knock-down sprays or combinations) show an emergence increase and/or following rains during emergence periods. Continue at 7-10 day intervals until emergence ceases.
Pecan Spittlebug	When 5% of terminals have spittle masses.

Treatment guidelines utilized for arthropod pests in 1977 are shown in Table 7. Since the primary purpose of pest management efforts in 1977 was to review proposed sampling procedures and action thresholds, absolute insect counts were not made each scouting day. Detailed counts were made only when pest levels were obviously near action threshold levels. Nevertheless, the program proved very effective. Only two of the six orchards under study required treatment for yellow aphids prior to August 1. These orchards required treatment in mid-May. The black pecan aphid was present in all orchards throughout most of the season but not in significant numbers. Both yellow aphids and black pecan aphids required control in all orchards in late August following the use of carbaryl for pecan weevil control. Aphid populations were reduced by addition of an aphicide to two of the pecan weevil applications.

During the season, pecan nut casebearers were found in four orchards and pecan spittlebugs in all orchards. Infestation levels of both pests remained below action thresholds throughout the season.

Pecan leaf scorch mites and their damage were found in one orchard during the latter part of May. Damage was confined to a few trees in a shaded area so treatment was not applied at that time. Populations remained low and the damage level did not increase until the middle of August. Mite populations were suppressed to acceptable levels in late August by the insecticide added to pecan weevil sprays for aphid control.

Significant numbers of pecan weevils began to emerge during the last week of July. Weevil control applications were initiated during the first week of August. This was approximately 2 weeks earlier than the normal starting date for weevil sprays. If weevil samplings techniques had not been used, significant damage would have occurred prior to initiation of weevil sprays. Once begun, weevil sprays were continued at 7-10 day intervals. Four applications were made to four orchards and five sprays were required in two orchards where weevil emergence was heaviest. One shortcoming in current weevil sampling techniques, of concern to the junior author, is failure to account for weevil distribution. Sampling techniques to account for aggregated weevil emergence have not been devised. Trap catches are only indicative of intra-orchard infestations *ie.*, weevil emergence in each orchard should be sampled. The feasibility of this approach is uncertain.

Shuckworm populations had not reached threshold levels prior to initiation of weevil sprays. After that time, hickory shuckworm counts were reduced to extremely low numbers.

The major benefit of the 1977 insect management program was the reduction of early-season insecticide applications (those aimed at yellow aphids and pecan nut casebearers). Due to scouting information on these pests, four insecticide applications were omitted. This saved participating growers ca. $17.00 per acre (insecticide cost). Improved control of other pests also resulted due to utilization of action thresholds. However, increased yields may have been confounded with other variables and therefore prevented accurate estimates of yield increases due to scouting. Thresholds for insects proved workable, but the need for further assessment of mite and weevil thresholds was apparent. Recent research (Boethel *et al.* 1976a, b; Eikenbary *et al.* 1977 and 1978) involving prediction of weevil populations based on trap catches offers promise for quantifying weevil thresholds. Hopefully, these techniques can be adapted to Georgia's growing conditions.

Disease Management. No specific action thresholds were utilized for disease control decisions. Scab infection of foliage and nuts was ranked according to degree of damage. Ratings ranged from 1 to 5, with 1 indicating no infection and 5 indicating very heavy or 100% infection (Phillips *et al.* 1952) as modified by Hunter and Roberts (1978). The rating system was valuable in determining relative abundance of scab and in evaluating the effectiveness of disease control practices. However, it was not used to make disease control decisions. Since there are no curative treatments for scab or other diseases after infection occurs, preventative fungicide applications were made. However, because of the extremely dry conditions, scouting did allow for extension of the time interval between mid-season fungicide applications. The decision to extend application intervals resulted in a reduction of 2 fungicide applications and a savings of ca. $13.00 per acre in material and application costs.

Another major benefit of scouting was derived from detection of diseases other than scab. Fungicide applicatons for pecan scab do not usually provide adequate control of diseases such as zonate leaf spot *(Cristulariella pyramidalis)* and powdery mildew *(Microsphaera alni)*. Low levels of each disease were observed in one orchard in 1977. Early detection of these sporadic, but potentially serious diseases, allowed time for control without yield losses. Early detection of minor disease problems is also important. An abbreviated fungicide spray schedule is recommended for scab tolerant varieties in Georgia (Ellis and Arnett 1978), and allows for development of minor diseases. Scouting of orchards has improved the value of this abbreviated program.

Summary of Results (1977). The sampling procedures used in our 1977 pecan insect management program accurately assessed pest populations in most instances. However, the number of samples taken in orchard surveys were increased as insect numbers approached action thresholds to gain confidence in treatment decisions. There was also concern over the feasibility of currently available weevil trapping techniques for large dispersed acreages. The action thresholds for insects proved workable although it was evident refinements would be necessary, especially for mite and weevil thresholds. Scouting also improved disease control practices by allowing accurate determination of scab levels and early detection of secondary diseases.

The pest management approach improved general efficiency of insect and disease controls and resulted in a savings of ca. $30.00 per acre by reducing pesticide input. Yields were also increased but quantifying these was not possible since yield responses were confounded with weather conditions and other factors.

1978 Program

In 1978, pest management efforts were expanded into a pilot, grower-supported field program. The program was conducted in Dougherty County under the leadership of county extension personnel. Two growers, 20 orchards and 1,433 acres were involved. One grower had 12 orchards totaling 930 acres and the second grower had eight orchards totaling 503 acres. The two locations were ca. 10 miles apart on opposite sides of the Flint River.

One scout was employed and trained prior to the season. Scouting was begun the first week in May, and continued at weekly intervals until September 21. Training was up-dated throughout the season by extension personnel. The cost of scouting was $4.00/acre.

It should be noted that pecan pest management training sessions were also conducted in three other counties. These were attended by growers, farm managers, commercial applicators and consultants. It is estimated that in 1978,

8,000 acres of pecans were under private or commercial pest management programs following our program guidelines. All have reported successful results.

The sampling procedures and action thresholds used in 1978 were the same as those used in 1977 with two exceptions: (1) The orchard survey procedure was modified to reduce scouting time by having scouts survey every fourth tree in every fifth tree row. Using this method, four trees per 20 trees (20%) were surveyed. This reduced sample size to two compound leaves and two nut clusters per tree instead of four. Counts were made from the top of a pick-up truck. A driver was furnished by one grower and drivers were hired for sampling other orchards. This survey procedure provided a high degree of accuracy in estimating pest levels due to more extensive sampling and allowed for a concomitant reduction in sampling time. (2) Due to widespread problems in 1977 with the pecan serpentine leafminer, an action threshold was established and utilized in 1978. This action threshold called for treatment when an average of one mine was found per leaflet and before more than five leaflets per compound leaf had mines.

Insect Management (Location 1). The major insect problems at Location 1 were yellow aphids, the pecan nut casebearer, pecan leaf scorch mite, and the hickory shuckworm.

Yellow aphids (primarily *M. nigropunctata*) were above the established threshold in all 12 orchards at Location 1 when scouting was initiated (Fig. 8). Treatments were not applied however, because aphid counts were declining. Three weeks later aphid counts increased and honeydew became very heavy. All 12 orhcards were treated at that time. Aphid populations did not exceed threshold levels again until mid-September. *M. caryella* was then the predominant species, although *M. nigropunctata* was also present. This is a normal pattern of occurrence (Tedders 1978a). Pecan leaf scorch mite populations also reached threatening levels in September, consequently an insecticide for mite suppression and aphid control was applied to all orchards (Fig. 8).

Infestations of pecan nut casebearer were found in all orchards in May. Infestations reached threshold levels in two orchards in late May, but the grower chose not to apply an insecticide. Second generation casebearers appeared in early July and caused significant nut drop from 'Stuart' and 'Pabst' varieties. Minor drop due to damage by second generation pecan nut casebearer occurred in all orchards but no treatments were applied. Subsequent generations were effectively suppressed by spray applications aimed at aphids and shuckworms.

Hickory shuckworms required two insecticide applications at Location 1. Relatively high populations appeared in June (Fig. 9) but there was no consistent increase in shuckworm catches in blacklight traps and the number caught in any trap never exceeded seven. Insecticide applications for shuckworm control were delayed until mid-July when threshold levels were reached. A

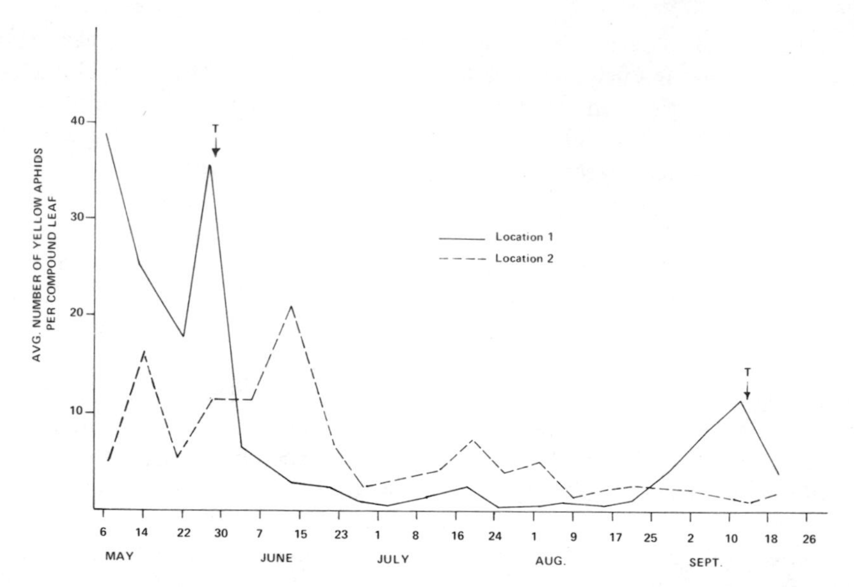

Figure 8. The infestation levels of yellow aphids at two locations in Dougherty County, Georgia throughout the 1978 growing season. T represents treatment dates.

second application was necessary during the third week of August when shuckworm populations once again reached the action threshold (Fig. 9). A material was used in this application that would also reduce late-season pecan nut casebearer populations. Black pecan aphids and leafminers were generally present throughout the season in low numbers but never reached action thresholds.

A total of four insecticide applications were made at Location 1. This was four less than used in conventional spray programs. In addition, more effective insect control was possible due to improved timing of applications.

Insect Management (Location 2). Before scouting was initiated, an insecticide application for yellow aphids had been made in all eight orchards at Location 2. Consequently, yellow aphid populations remained below action levels until mid-June (Fig. 8) when populations exceeded action threshold levels in three orchards. Aphid populations declined without treatment and were held at acceptable levels by applications made for mites.

Hickory shuckworm levels were fairly high in late June and early July (Fig. 9) but did not exceed established threshold levels. Late-season populations were controlled by applicatons aimed primarily at mites.

Pecan leaf scorch mites were the major problem at Location 2. Low mite levels and light damage were noted in two orchards in mid-June. Significant levels of mites and damage were present in all orchards by mid-July, when mite

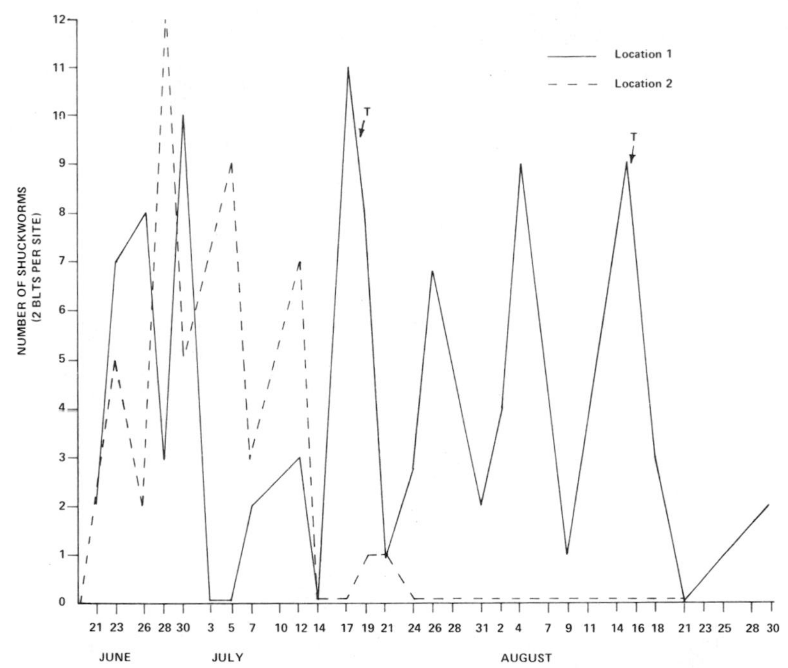

Figure 9. The number of adult hickory shuckworms captured in blacklight traps at two locations in Dougherty County, Georgia in 1978. T represents treatment dates.

controls were applied. A second application was made seven days later. Mites reached damaging levels again in mid-August and persisted. Suppression of mites required addition of an insecticide to each fungicide application made throughout the remainder of the season. A material was selected that provided both mite suppression and control of other late-season pests. These applications kept population levels of the black pecan aphid, yellow aphids (Fig. 8) and hickory shuckworm (Fig. 9) suppressed below their respective action thresholds throughout the remainder of the season.

Six orchards at Location 2 received five insecticide applications and two orchards received seven applications. All applications, except one made in each orchard for aphid control, were aimed at scorch mite suppression. In spite of high mite populations, the program reduced insecticide input by 2.5 applications when compared to conventional spray programs used at this location. This resulted in a savings of ca. $10.87 per acre.

All orchards in the 1978 pest management program were free of pecan weevils. However, in a related program, cone cage traps were used to monitor weevil emergence (Fig. 10) in 12 South and Middle Georgia counties to provide growers with information on the status of weevil emergence in their general

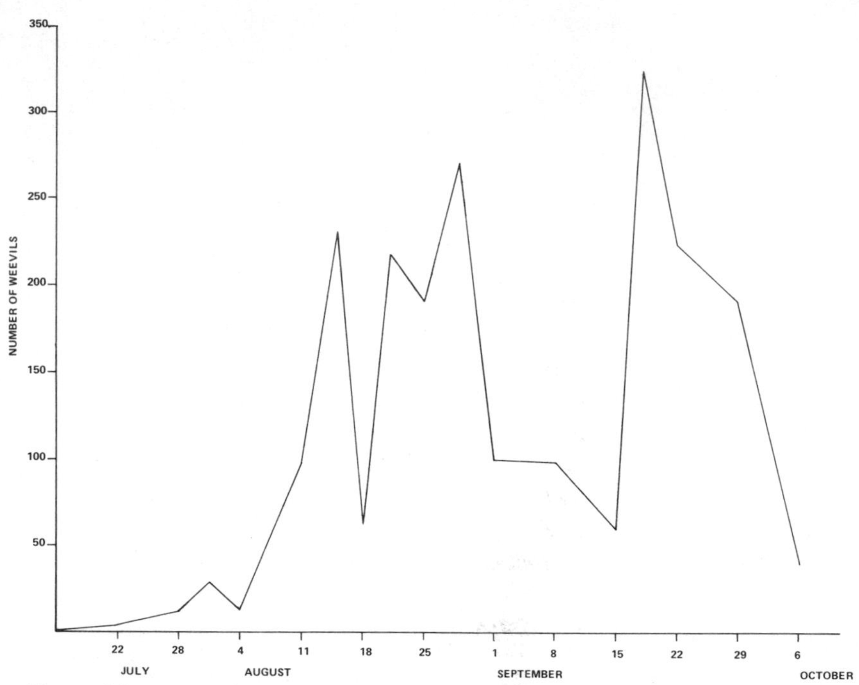

Figure 10. Seasonal occurrence of the pecan weevil in 12 Georgia counties in 1978.

areas. Extension service, experiment station and USDA personnel cooperated in installation and checking of the traps. From this program, information regarding weevil emergence was relayed to county extension offices and ultimately to growers through telephone calls and newsletters. The program provided many growers with information to improve timing of weevil sprays.

Disease Management. Preventive fungicide spray schedules were followed at both locations in the 1978 pest management program. An abbreviated fungicide schedule was followed on scab tolerant varieties but there was no program-related reduction in the number of fungicide applications. Good protection from pecan scab was obtained in all orchards.

High infection levels of downy spot *(Mycospharella caryigena)* were detected on trees of the 'Stuart' variety in most orchards at Location 1. Preventive treatments for this disease must be included in the "bud-break" spray. The detection of downy spot by the scouting program will provide proper preventive measures in the 1979 production program.

Summary of Findings (1978). The 1978 scouting program worked well, although the scout was relatively inexperienced with pecan pests. Scouting proved to be an effective management tool for pecan insect pests and a useful tool to detect the need for temporary provisions in disease preventive spray programs.

The action thresholds again proved workable but, future management programs will allocate more time for attention to second generation pecan nut casebearers.

It is estimated that, using the stated procedures, one scout can effectively survey from 2,000-3,000 acres in high density pecan areas. Less acreage could be surveyed in areas with smaller, scattered orchards. The $4.00 per acre scout cost was adequate in 1978. Scouting prices would necessarily increase if extended travel were involved or if additional services were offered.

The pest management program is a highly desirable replacement for conventional spray programs. It improved the timing and consequently the efficacy of pesticide applications, and generally reduced production costs. To pecan growers, this program offers great potential benefits which will increase as the program is refined, expanded and strengthened.

ACKNOWLEDGMENTS

Grateful appreciation is extended by the junior author to the various state and USDA pecan researchers whose work is a part of the Georgia program. Special thanks are extended to J. A. Payne and J. D. Dutcher for their assistance in sampling pecan weevils, to W. L. Tedders for conducting a hickory shuckworm identification clinic, and to J. S. Smith for use of blacklight traps.

REFERENCES

Boethel, D. J., R. D. Morrison, and R. D. Eikenbary. 1976a. Pecan weevil, *Curculio caryae* (Coleoptera: Curculionidae): 2. Estimation of adult populations. *Can. Entomol.* **108**:19-22.

Boethel, D. J., R. D. Eikenbary, R. D. Morrison, and J. T. Criswell. 1976b. Pecan weevil, *Curculio caryae* (Coleoptera: Curculionidae): 1. Comparison of adult sampling techniques. *Can. Entomol.* **108**:11-18.

Eikenbary, R. D., W. G. Grovenburg, G. H. Hedger, and R. D. Morrison. 1977. Modification and further evaluation of an equation for predicting populations of *Curculio caryae* (Coleoptera: Curculionidae). *Can. Entomol.* **109**:1159-1166.

Eikenbary, R. D., R. D. Morrison, G. H. Hedger, and D. B. Grovenburg. 1978. Development and validation of prediction equations for estimation and control of pecan weevil populations. *Environ. Entomol.* 7:113-120.

Ellis, H. C., and J. D. Arnett. 1978. Pecan insects and diseases and their control. *Ga. Ext. Ser. Bull.* 644. 20 pp.

Hunter, R. E., and D. D. Roberts. 1978. A disease grading system for pecan scab. *Pecan Quarterly* **12**:3-6.

McVay, J. R., G. R. Strother, W. S. Gazaway, R. H. Walker, and J. Boutwell. 1977. The Alabama pecan pest management program. *Ala. Coop. Ext. Ser. Rept.* 32 pp.

Nash, R. F., and C. A. Thomas. 1972. Adult pecan weevil emergence in the upper coastal plains of South Carolina. *J. Econ. Entomol.* **65**:908.

Phillips, A. M., J. R. Cole, and J. R. Large. 1952. Insects and diseases of the pecan in Florida. *Fla. Agr. Exp. Sta. Bull.* 499. 76 pp.

Polles, S. G., and J. A. Payne. 1972. An improved emergence trap for adult pecan weevils. *J. Econ. Entomol.* **65**:1529.

Polles, S. G., and J. A. Payne. 1973. Techniques for timing spray application to control the pecan weevil. *Proc. Southeastern Pecan Growers Assoc.* **66**: 101-108.

Raney, H. G., and R. D. Eikenbary. 1969. A simplified trap for collecting adult pecan weevils. *J. Econ. Entomol.* **62**:722-723.

Tedders, W. L. 1973. A new method of determining the emergence of pecan weevils. *Proc. Southeastern Pecan Growers Assoc.* **66**:111.

Tedders, W. L. 1974. Bands detect weevils. *Pecan Quarterly* **8**:24-25.

Tedders, W. L. 1976. Beneficial insects of pecans. *Pecan South.* **3(4)**:400-402.

Tedders, W. L. 1978a. Important biological and morphological characteristics of the foliar feeding aphids of pecan. *USDA Tech. Bull.* No. 1579. 29 pp.

Tedders, W. L. 1978b. Personal communication to J. R. McVay.

Tedders, W. L., and M. Osburn. 1966. Blacklight traps for timing insecticide control of pecan insects. *Proc. Southeastern Pecan Growers Assoc.* **59**: 102-104.

Tedders, W. L., J. G. Hartstock, and M. Osburn. 1972. Suppression of hickory shuckworm in a pecan orchard with blacklight traps. *J. Econ. Entomol.* **65**:148-155.

West, R. P., and M. Shepard. 1974. A modified cone emergence trap with increased capabilities for pecan weevils. *Fla. Entomol.* **57**:357-360.

INTEGRATED PEST MANAGEMENT OF INSECTS AND MITES OF PEAR

P. H. Westigard
Oregon State University
Southern Oregon Experiment Station
Medford, Oregon

INTRODUCTION

The three Pacific coast states of California, Oregon and Washington produce in excess of 90% of the commercially grown pears in the United States. Within this geographically limited area, there exist quantitative and qualitative differences in pest fauna. These differences along with substantial variation in climate, cultivar dominance, and horticultural practices demand that the data base necessary for the implementation of integrated pest management (IPM) programs be developed to meet the requirements and needs of individual pear districts. This chapter deals with the development of IPM of insect and mite pests in one such pear district, southern Oregon.

Since about 1900, when the pear industry became firmly established in this area, losses due to arthropod pests have been an important and occasionally limiting factor to the continuous production of high quality fruit. As is true of other agricultural crops, the pest control tactics utilized by pear growers have passed through distinct stages. In southern Oregon these control stages can be categorized as (1) utilization of marginally effective inorganic pesticides combined with labor intensive cultural control methods (*ca.* 1900-1945), and (2) nearly sole reliance upon synthetic organic pesticides applied on a calendar or preventative schedule (1945-present).

Both stages have had associated problems, primarily sub-economic control (primarily stage 1), environmental contamination (stage 1 and 2), illegal chemical residues (primarily stage 1), pesticide resistance (stage 2) and destruction of beneficial species and subsequent resurgence of non-target secondary pests (primarily stage 2).

Over the years these problems have periodically threatened the survival of the pear industry either directly through lack of control, caused by resistance or to the absence of effective chemicals, or indirectly by pesticide legislation attempting to reduce the threats to user or consumer safety or to the environment. IPM is an attempt to resolve the contradictory demands of a society that on the one hand asks for a constant supply of high quality fruit free of pest injury, but on the other is concerned with the potential negative side effects of pesticide usage. The pest management approach attempts to utilize an array of suitable control techniques rather than relying on a single disruptive tactic. The tactics suggested include the use of biological control agents (predators, parasites and pathogens), host plant resistance, cultural controls, pesticides, and behavioral controls. Summarizing the studies conducted in southern Oregon over the past 15 years I will attempt to show the applicability of the various control tactics for each major pest.

Another important difference between IPM and the stage 2 approach to pest control lies in the acceptance of sub-economic pest densities under IPM, rather than demanding nearly total elimination as a criterion for successful control. This idea is most basic to integrated pest management as, for example, the utilization of biological control agents as a viable control tactic depends upon residual pest levels to insure the survival of natural enemies.

The above concept, however, demands that a new type of data base be developed before any control tactic is chosen, namely the establishment of an economic injury level and of an economic injury threshold for pest density. The economic injury level has been defined by Stern *et al.* (1959) as "the lowest population density that will cause economic damage. Economic damage is the amount of injury that will justify the cost of artificial control measures..." These same authors define economic threshold as "the density at which control measures should be determined to prevent an increasing pest population from reaching the economic injury level." The establishment of economic injury or economic threshold densities is, however, not an easy calculation as crop value varies dramatically from year to year, orchard to orchard and variety to variety. Estimates of these values are given in the sections dealing with individual pest species and in the section on economic aspects of IPM.

Another requirement for IPM that relates to economic injury thresholds and pest damage levels is that of sampling. Sampling of pests and their natural enemies in an IPM program requires a methodology that relates pest density to potential economic loss. With a high value crop such as pears, the economic tolerance for injury is quite naturally low and, given the variability encountered in pest distribution, the required sample size is often large and sometimes economically prohibitive. Sampling procedures then are another critical aspect of IPM implememtation and are discussed for each major pest.

In addition to the entomological aspects of integrated pest management, there are important non-entomological inputs which determine to a great extent the feasibility of IPM implementation. These include horticultural and economic considerations, which are discussed as limited data and background permit.

The scope of this study is purposefully limited to the development of a data base to promote the implementation of an IPM program for pears in southern Oregon. Our principal sources of information are drawn from studies conducted in the area between 1963 and 1978. In some cases we have used data from other geographic areas where the information has been tested and appears to fit southern Oregon conditions. In addition to the specific references cited we have made use of several review articles, especially those by Hoyt and Burts (1974), Madsen and Morgan (1970) and Barnes (1959), which deal with integrated pest control in orchard ecosystems.

CODLING MOTH

The codling moth *Laspcyrcsia pomonella* (L.), has been a key pest in pear orchards since the earliest days of pome fruit production in southern Oregon's Rogue River Valley. Records from the area indicate that between 1916 and 1945 despite chemical control programs, the damage attributed to this species resulted in losses of 5 to 30% of the pear crop annually (Cordy 1977). During this early period the complete program included six or seven summer sprays of inorganic inscctícides which were applied by hand using high pressure equipment.

When DDT was first introduced in 1946 the results were nothing short of spectacular, reducing the level of wormy fruit to near zero with only three to four applications. This low infestation level has, with few exceptions, been maintained to the present day. However, the codling moth, though presently well-controlled, still remains the most important arthropod pest in southern Oregon pear orchards.

Life History

The codling moth can be considered to be oligophagous as it attacks a variety of tree crops ranging from nut crops to stone and pome fruit crops.

The life history of the pest is similar on all hosts. The species overwinters as a diapausing mature larva and is found predominately in cracks or crevices or under loose bark on the major scaffold limbs or trunk of the host tree. Some overwintering larvae may also be found in soil in close proximity to the tree crown.

With diapause termination in late winter the mature fifth instar larvae pupate and, depending upon temperature, adults begin to emerge in early spring. First adult flight in southern Oregon usually occurs in mid to late April. The flight pattern of the adult is dictated by abiotic factors including temperature. California workers have found that moth flight is restricted by temperatures below $55^{o}F$ or above $80^{o}F$ (Batiste *et al.* 1973). These parameters also seem to fit conditions in southern Oregon and are important considerations in the interpretation of adult monitoring devices used to time the application of control tactics directed toward this stage.

Following emergence and mating, the female moth begins to deposit eggs on or near fruit. Our observations indicate that the presence of fruit is a necessary stimulus for oviposition. Few eggs are laid in harvested orchards or in orchards which have lost their fruit due to spring frosts. Eggs are laid singly on leaves or fruit and pass through distinct developmental stages. When first laid, eggs are translucent but change in appearance as the embryo develops. On pears the majority of eggs are laid on the underside of leaves with decreasing numbers laid on fruit, fruit stems or on the upper leaf surface (Westigard *et al.* 1975).

Upon hatching, the first instar larvae will begin a pattern of search for a suitable fruit entry site. On the pear host there appears to be a significant difference between varieties in the success of larvae to enter fruit. Upon successful fruit entry, the larva will generally penetrate to the seed cavity. After completion of four molts the larva is mature and will generally leave the fruit to seek a pupation site. Under southern Oregon conditions there are usually two complete generations a year. A partial 3rd generation has been reported in the area on apples (Yothers and VanLeeuwen 1931).

Since temperature is an important parameter describing codling moth development, researchers in California (Pickel 1976) and Michigan (Riedl *et al.* 1976) have attempted to construct a predictive model to relate codling moth phenology to temperature regimes. These models have been tested under southern Oregon conditions and appear to offer some promise in predicting important phenological events (Table 1).

Damage

The damage caused by the codling moth is inflicted by the larva in its penetration into the pear flesh. These entries are not only unattractive to

Table 1.

Phenological events of the codling moth in 1976 as predicted by temperature models compared to observed events under southern Oregon conditions.

Event	Dates Predicted		Dates Observed
	Univ. Calif. Model	Michigan State	
1st oviposition	April 30	April 30	May 12[a]
1st egg hatch	May 12	May 15	May 13
1st larval spin in bands	June 22	July 4	June 30
1st emergence summer moths	July 10	—	July 19

[a]The first egg was found in 'black head' or late stage of development indicating oviposition having occurred some time prior to the date first observed.

consumers but also cause problems in sorting and storage of fruit. The allowable damage by the codling moth as determined by the various grade standards set for the fresh fruit market is nearly zero. In essence, fruit infested with codling moth damage to offset present control costs is in the range of 1% fruit loss.

Sampling

Several types of population monitoring have been used for adult codling moths such as the use of fermenting bait traps, blacklight traps and sex pheromone baited traps. These devices have been used primarily as aids in the timing of pesticide treatments. Recently the pheromone traps have been shown to be useful in determining the need for treatment based on estimates of population density.

Adult sampling for timing. The proper timing of treatment will depend upon the type of activity exhibited by the control agent along with the availability and sensitivity of sampling devices to determine the pest stages present. In the case of the codling moth, the adult, egg and pupal stages represent non-injurious forms which if properly monitored can forecast the proper treatment timing prior to injury expression. Historically, the adult stage of the codling moth has been used in this manner, as the field detection of eggs is time consuming, especially at low density levels.

Current treatment timing in southern Oregon usually takes the form of detecting, with pheromone traps, the time of the first substantial flight of the overwintering and of the first summer generation male moths. When these are

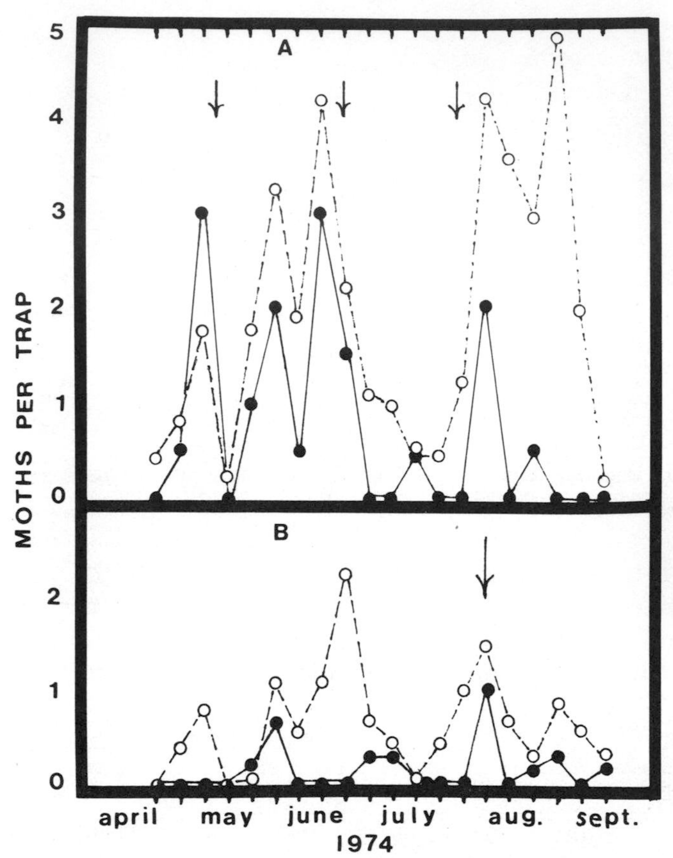

Figure 1. Codling moth catches in sex pheromone baited traps in two southern Oregon pear orchards. Arrows indicate dates of pesticide application. Dotted lines indicate moths captured in border traps, solid lines moths in interior traps. [Modified from: Westigard and Graves (1976)].

detected a pesticide treatment is usually recommended (Fig. 1). The emergence of overwintering adults is of about a two month duration, and with a 2-3 week residual effect of available pesticides, an additional spray is routinely applied about 30 days following the initial treatment. Depending upon the harvest period of pear variety involved, an additional 1 or 2 sprays are applied for the first generation summer moths which begin to appear in mid to late July.

Adult sampling for treatment need. Though the pheromone traps have been useful in proper pesticide timing, another contribution to IPM is made by relating the magnitude of moth catch to potential larval damage and thereby forecast the need for treatment (Fig. 1). As will be discussed in other sections, a reduction in the number of pesticide sprays needed for control of the codling moth will enhance the survival of several important natural enemies of other pear pests.

Preliminary work in southern Oregon to evaluate the relationship between pheromone trap catches of male moths and damage followed that conducted by Madsen and Vakenti (1973). In their tests, one trap was used to monitor about two acres of orchard area and if catches exceeded 2 moths/trap/week, a spray was recommended. In order to adjust for non-entomological grower practices in southern Oregon, such as irrigation schedules, and for differences in varietal susceptibility to codling moth damage, modifications have been made in the codling moth monitoring program. Studies showed that economic damage was avoided on the 'Bartlett' variety by postponing sprays until an average of 10 moths/interior trap had been accumulated. On the 'D'Anjou' variety, it appeared a level of 20 moths/trap might be tolerated (Westigard 1978). During these studies the total number of moths caught was accumulated until the above levels were reached at which time a chemical was applied. Following treatment, moth accumulation was begun after 14 days. This interval represents the approximate period of time that toxic chemical residues would be present on fruit and foliage; therefore, moths caught during this time would not represent an economic threat. The current tentative criteria for treatment levels for the codling moth are given in Table 2.

Control Tactics

Biological control. In the absence of chemical treatments the codling moth is attacked by a variety of biological agents. These include egg, larval and pupal parasites and predators as well as disease organisms. Unfortunately these are unable to maintain the species under a commercially acceptable degree of control. Studies in an unsprayed 'Bartlett' pear orchard from 1964-1971 showed that the codling moth infestation in this untreated situation ranged from 22 to 81% and averaged 42% (Westigard 1973b). In a portion of these tests the mortality to eggs and early larval instars was assessed. Egg mortality averaging 26% was measured and attributed primarily to the activity of parasites and predators (Table 3). Mortality to the first instar larvae averaged 55% which was though to be primarily due to host plant resistance (antibiosis). The natural mortality to other codling moth stages was not measured (Westigard *et al.* 1975).

Chemical control. As described previously, the natural controls operating on the codling moth are unable to maintain population density at a level below the current injury threshold of about 1%. Therefore, this pest has required the use of artificial controls, namely insecticides. From the early 1900's to the mid 1940's growers relied upon inorganic chemicals to achieve control and from the 1940's to the present upon synthetic organic pesticides. Both types of compounds have produced problems related to the side effects of their use.

Table 2.

Tentative criteria for the need to apply chemical treatment for economic suppression of the codling moth.

Period	♂ moths accumulated from	Accumulated moth catch/interior trap Variety Bartlett	 D'Anjou	Treatment
1st spray period *ca.* May 10-20	1st ♂ catch	>10 <10	>20 <20	yes no
2nd spray period *ca.* June 10-20				
A. treated 1st spray period	14 days after 1st spray	>10 <10	>20 <20	yes no
B. not treated 1st spray period	1st ♂ catch	>10 <10	>20 <20	yes no
3rd spray period *ca.* July 20-Aug. 10				
A. treated 1st or 2nd spray period	14 days after last spray	>10 <10	>10 <10	yes no
B. not treated 1st or 2nd spray period	1st ♂ catch	>10 <10	>10 <10	yes no

From the early 1900's to the introduction of DDT in 1945, arsenical insecticides were the principal chemicals used for codling moth suppression. For a short period, 1943-1945, cryolite was used on a limited basis. The problems associated with use of the arsenicals included relatively poor control (infestations of 10% were not uncommon in treated orchards), excess residues on fruit (the growers developed a special fruit wash to remove arsenic residues) and the buildup of toxic residues of these compounds in the soil.

DDT was the only chloronated hydrocarbon insecticide used for codling moth control in southern Oregon. This compound was used for the first time in 1945 and by 1947 had essentially replaced the inorganic compounds. Because of DDT's persistance and mode of action the number of applications required to obtain control was reduced from 6-8 using inorganic chemicals to 2-4. In addition, DDT also reduced the amount of codling moth damage to nearly zero compared to the 5-30% losses experienced with the previously used inorganic compounds.

No significant problems with DDT residues on pear fruit were experienced in southern Oregon during its period of use (1947-1960). Primarily this was due to the lowered number of applications necessary to achieve control and to the

Table 3.

Field mortality of egg and 1st instar larvae of codling moth on 'Bartlett' pear. 1970-1971. Medford, Oregon.[a]

Year	Eggs					1st instar larvae			Total egg & larval mortality
	No. found	No. parasitized	No. shrivelled	No. disappeared	% mortality	No. entering fruit	No. lost	% mortality	
1970	85	7	9	1	20	16	52	76	81
1971	90	13	8	9	33	39	21	35	57
$\bar{X}$	87.5	10	8.5	5	26.5	27.5	36.5	55	69

[a]Taken from: Westigard *et al.* (1975).

relatively high residue by law (7 ppm). However, buildup of DDT and its metabolites in the soil has been of concern. In 1960 it was estimated that nearly one half of the DDT used previously had been accumulated in the orchard soil. By 1965, 5 years after the discontinuance of DDT in pear orchards the residue had decreased by only 50% (Terriere *et al.* 1966).

It is somewhat difficult to assess the total effect of DDT on non-target species in southern Oregon pear orchards. On the one hand, the increase in spider mite levels was dramatic with the twospotted spider mite becoming an annual rather than an occasional pest. On the other hand the incidence of damage due to so-called minor pests such as fruit tree leafroller, pear thrips and green fruit worms decreased in severity.

Only two organophosphate compounds, azinphosmethyl and phosmet, have been widely used in southern Oregon for codling moth control. As was the case for DDT, both phosphate compounds have produced negative as well as beneficial effects in the management of pear pests. First introduced in the late 1950's, azinphosmethyl gave excellent control of the codling moth and of the then recently introduced pest, the pear psylla. No problems have been experienced in the accumulation of excess residues on harvested pears nor to resistance by the codling moth to the phosphate pesticides.

The two phosphate insecticides used for codling moth control have been useful in the establishment of an integrated control program for spider mites on pear. This program developed as a result of the selectivity exhibited by azinphosmethyl in providing control of the codling moth but not eliminating the important spider mite predator, *Metaseiulus occidentalis* Nesbitt. The organophosphates have also been effective in the suppression of several non-target phytophagous species.

On the negative side, the use of the organophosphates has caused the destruction of a great many natural enemies resulting in increased density of the pear psylla. This has become especially evident with the failure of these chemicals to provide direct psylla suppression.

With the appearance of resistance by the pear psylla to phosphate pesticides in the early 1970's, southern Oregon pear growers shifted to chlordimeform as the material of choice for control of pear psylla and the codling moth. Chlordimeform is highly toxic to predaceous mites and with the widespread use of this chemical beginning in 1970, the integrated program using *M. occidentalis* to achieve commercial suppression of spider mites was essentially destroyed. In addition, the exclusive use of this compound during the summer months led to increased damage from the San Jose scale, a pest which was apparently suppressed by the organophosphate materials previously used.

Host plant resistance. In the section dealing with biological control of codling moth it was shown through the use of a modified life table that about 70% of the eggs and 1st instar larvae do not reach maturity. Of this, more than

75% was attributed to 1st instar larval mortality. In subsequent studies it was suggested that this mortality was due to the inability of the young larva to enter the pear. It was found that this was not only greater on pear than on apple but also varied with the pear variety being attacked and with time. Examination of Fig. 2 will show that of the pear varieties tested, the 'Bartlett' variety was the most susceptible, followed by 'Bosc', 'Comice', and 'D'Anjou'. The phenology of resistance shows that a period of relatively moderate tolerance is followed by a longer period of high tolerance and then by a period of high susceptability. The difference in host suitability is currently explained by the formation of "stone cells" below the epidermal layer. This process, referred to as lignification, usually occurs in early June in southern Oregon. Stone cells are then present until the ripening process is initiated. This would account for the early season susceptibility of all varieties, followed by decreased ability of larvae to enter fruit and finally by increased susceptibility toward harvest period.

From an IPM standpoint, the above findings have allowed the establishment of a higher economic injury threshold for the 'D'Anjou' variety and offer the potential for lowering the number of chemical treatments required on this winter variety (Table 2).

Cultural control. As mentioned in the introduction there are several non-entomological considerations which bear on the potential for implementation of an IPM program. Those that are most pertinent to codling moth control in southern Oregon include irrigation scheduling, overtree irrigation, and orchard sanitation practices.

The orchard land in southern Oregon requires summer irrigation in order to maintain tree vigor and to properly (commercially) size the pear fruit. Typically growers will apply four and possibly five flood irrigations between early June and late August. Because of heavy soil types in the area, once an irrigation has begun it is virtually impossible for heavy orchard equipment, such as air blast sprayers, to pass through for 14-21 days. This prohibits to some extent curative spray treatments and forces growers ro rely upon the more disruptive practice of preventative spray scheduling. Pest control decision making periods during the late spring and summer months are therefore limited and once made, must provide predictable and commercially acceptable control for a minimum 3-week period. In addition, irrigation scheduling makes it difficult to utilize some of the important, finely timed pest management tools such as the codling moth temperature-phenology models which allow for more precise timing of pest control procedures.

A major limiting factor in the production of pears in southern Oregon is the nearly annual occurrence of severe spring frosts. In order to minimize frost damage growers have employed orchard oil heaters, wind machines, and recently overtree sprinklers (Lombard *et al.* 1966). The latter type of frost prevention is a

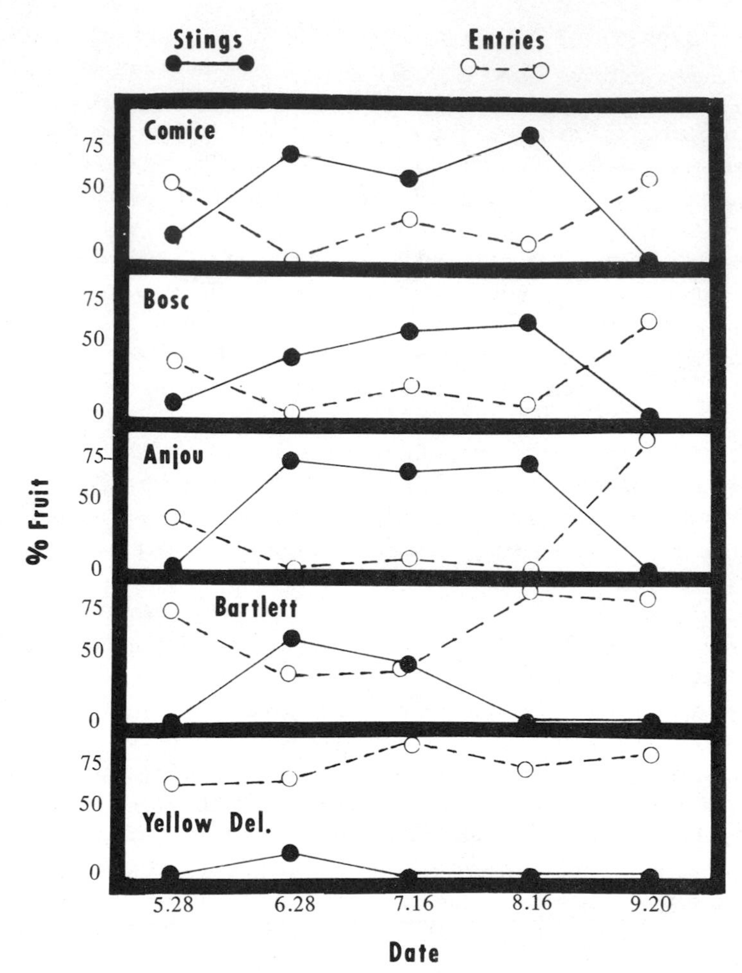

Figure 2. Relative susceptability of pear and apple varieties to entry by first instar codling moth larvae. [Taken from: Westigard *et al.* (1975)].

multi-purpose installation also used for summer irrigation and tree cooling. Overtree irrigation in summer months has been found to have deleterious effects on pesticide residues, including those needed for codling moth control. A partial solution to this problem was found by reversing the normal spray-then-irrigate pattern. This in turn demanded that other orchard practices be modified. That is, in place of clean cultivation, orchard rows were planted to sod, which made possible the use of orchard spray equipment soon after an irrigation (Westigard *et al.* 1974).

Fig. 1 presented the codling moth catches in pheromone traps in several Rogue Valley orchards. These catches are divided between those found in pheromone traps placed in the inside of the orchards and those found in traps placed around the orchard periphery. The border traps captured nearly 80% of the moths recorded. This situation has been typical of most orchards monitored in southern Oregon and is thought to result from the movement of codling moths from abandoned pear trees into commercial orchards. In the early 1970's, a program of abandoned tree removal was initiated by southern Oregon pear growers and if successful will reduce the pest innoculum harbored in these abandoned blocks.

SPIDER MITES

There are four species of tetranychid mites that are pests of pear in southern Oregon. These are the twospotted, *Tetranychus urticae* Koch; European red, *Panonychus ulmi* (Koch); yellow, *Eotetranychus carpini borealis* (Ewing); and McDaniel, *Tetranychus mcdanieli* (McGregor) spider mites. Of these, the twospotted spider mite is the most widespread and the most important economically. Efforts to build an integrated pest management program for spider mites have dealt predominately with *T. urticae* and indepth data concerning the other species is lacking.

The twospotted spider mite was first reported as a pest of pears in 1924 but formal identification was not made until 1937. From that time until the mid 1940's this species caused sporadic but occasionally severe damage. With the introduction of synthetic organic pesticides in 1945, and to the present time, the twospotted mite can be classified as an annual pest that now receives 2-3 summer acaricide treatments in most commercial orchards.

The second most common and destructive spider mite is the European red mite which was not reported from southern Oregon pear orchards until 1953. Presently, it is controlled by an application of petroleum oil in the dormant period and by summer sprays primarily directed toward the twospotted mite.

The other two species of spider mites are found only in a few scattered orchards and only during some years. The identification of the yellow mite occurred in 1938 and the McDaniel mite in 1972. Commercial control measures for these species rely upon acaricides, usually the same as those used for the twospotted mite.

Life History

Though the life histories of these tetranychid mites are similar, several differences exist which are important to their management. These include variations in biological attributes which determine (1) suitable timings for control tactics, especially chemical controls; (2) reproductive potential which influences damage potential; (3) intra- or inter-tree distribution as it bears on sampling procedures; and (4) behavioral differences especially those dealing with species vagility.

Except for the European red mite, the spider mites attacking pear overwinter as diapausing adult females. These forms are found in cracks, crevices, and behind bark flakes on the trunk, on major scaffolds or on smaller limbs. Depending upon temperature, the three species move from overwintering sites about late March or early April. This period coincides with the opening of pear buds which become the initial feeding sites. Though chemical treatments are applied during the dormant or delayed dormant period for control of other pests, these treatments rarely have depressing effects on populations of two-spotted, yellow, and McDaniel spider mites.

The European red mite overwinters in the egg stage, exposed on fruit spurs and on smaller, newer pear wood. Thus, chemical treatment can be applied with success against this species in the prebloom period. Hatching of the overwintering eggs occurs about mid-April and normally coincides with the bloom period of the various pear varieties. As the case with the other spider mite species, the initial feeding area is on leaves in fruit clusters. From the time of habitation of the fruit cluster and until overwintering begins in the early fall, all spider mite species are to be found feeding on leaf tissue in various locations on the tree.

Reproductive Potential. The reproduction potential of spider mites is quite variable even under controlled conditions. For instance, laboratory tests conducted in 1969 compared the fecundity of the yellow mite with that of the twospotted mite. There were nearly six times the number of eggs laid by the twospotted mite compared to the yellow mite (Westigard and Berry 1970). Though we have not conducted studies with other spider mite species, a similar degree of variation would be expected. In addition to the innate rate of increase there are several abiotic factors that influence the rate of population growth. Of these, temperature is probably the most important. Since temperature is an important input into population density potential, it becomes a prominent aspect of pest management not only for the spider mite group but for other pest species as well. Unfortunately for the practice of pest management, the day to day temperature changes that influence pest levels are not accurately predictable and must be dealt with in a hypothetical manner. This 'what if?' interpretation of pest potential generally leads to over use of pesticides. However, until more accurate long-term weather forecasts are available, it may benefit pest management consultants to make decisions as to appropriate control tactics based on reasonably expected extremes rather than on average temperature accumulated over several years.

Intra-tree distribution. Knowledge of the intra-tree distribution of mite species on pear is important to the development of sampling procedures usuable in an IPM program. Comparison of the dispersion of the yellow mite and the twospotted spider mite was made on mature 'D'Anjou' trees. These data show that despite a high degree of species overlap there were differences in both vertical and horizontal planes. In general the twospotted mites were more commonly found on leaves in close proximity to the major scaffold limbs of the tree, while the yellow mites were found more frequently on leaves on smaller limbs (Westigard and Berry 1970).

Dispersion. As previously described, the movements of the codling moth from outside sources into commercial orchards are important to the practice of IPM. Immigration of spider mites have also been noted, not only by passive movement of adults but also by dispersion from ground litter or cover crops. Data from southern Oregon shows that about 50% of the population density of the twospotted mite on pear trees may be attributed to these upward movements from ground cover sources. Wind drift of twospotted mites has also been recorded but the contribution of this factor to eventual spider mite density on pear has not been measured (Westigard *et al.* 1967).

In terms of pest management, the phenomena of spider mite dispersion is capable of dramatically altering the expected mite density calculated from consideration of initial pest levels, reproductive capacity and temperature regimes. Thus, with the possibility of substantial dispersion occurring, spider mite levels must be continually and frequently monitored in order to avoid underestimates of population density.

Damage

The spider mites attacking pear are considered to be indirect pests. That is, they inflict damage to leaf photosynthate that in turn is reflected in effects on fruit or on other aspects of tree productivity. However, at times spider mites may affect pear quality directly by feeding on the epidermis of the pear fruit which causes russetting and subsequent downgrading.

As is true with other pear pests, spider mite density determines actual damage to the pear crop. Tests to measure the relationship of mite density to damage demonstrated that twospotted mites influenced several aspects of pear tree productivity and that the degree of damage was density dependent. Of the tree responses evaluated, fruit set (the number of fruit to reach maturity per 100 fruit clusters) was the most responsive to mite feeding. This effect, however, was expressed the year following mite damage (Table 4). Fruit size was also affected by mite feeding but the only consistant differences were those measured between the completely untreated plot and the plot that was kept below 5 mites/leaf. Increases in preharvest fruit drop and lower fruit grade were also noted in untreated plots when compared to the 5 mite/leaf level.

Table 4.
Influence of 1964 mite levels on fruit set and yield of 'D'Anjou' pear trees in 1965.[a]

1964 treatment level	1965[b]	
	Fruit set (fruit/100 clusters)	Yield (40-lb. lugs)
Mites/leaf	No.	No.
5	19.7 a	18.5 a
10	18.4 ab	18.1 a
25	16.0 ab	17.8 a
Untreated	10.8 b	7.5 a

[a]Taken from: Westigard *et al.* (1966).
[b]Means followed by the same letter are not significantly different.

Another aspect of spider mite damage to pear is the relationship between the time of attack and the potential crop loss. Again, using the twospotted mite on the 'D'Anjou' host, studies were conducted to measure this relationship (Westigard *et al.* 1967). Seven time intervals were chosen during which mites were allowed to feed and reproduce unchecked. Preceeding or following these periods the trees were kept as mite free as possible. The mite densities in each plot and the effect on fruit quality, size and tree yields were measured. It was found that fruit finish was influenced by late season feeding while fruit size was more influenced by early summer injury. No relationship was found between fruit set characteristics and a particular time of mite abundance. Rather, it appeared that plots with the greatest summer mite densities, regardless of time of attack, were those in which the following year's fruit set and therefore yield was the lowest.

Sampling

Sampling of spider mite populations is usually begun after bloom and continued through the late spring and summer months. These samples may be used to determine presence or absence of the various spider mite species and to calculate their density levels. Spider mite detection is important because of differences between species in susceptibility to various chemical acaricides.

Estimates of spider mite density are used to determine the need for treatment and to evaluate previously utilized controls. The sample unit used to make estimates of summer mite density is the mature pear leaf. Using this unit, a sampling methodology has been developed for use in commercial pear orchards of southern Oregon. In this system, five mature pear leaves are taken from each sample tree. Because of the differences between mite species in their intra-tree distribution, one leaf is taken from five locations along a single major scaffold limb beginning near the tip and extending to near the trunk area. The number of trees to be sampled is dependent upon intra-tree mite density variation and upon the accuracy required. Between tree variation in mite density appears to be dependent on population density and the mite species involved. Studies have shown that the between tree variations were inversely related to population density of the twospotted mite and the European red mite, and for the important predator mite, *M. occidentalis* (Westigard and Calvin 1971).

The values obtained in the above study were used to calculate the number of trees to be sampled to achieve the desired accuracy. These estimates of sample size (# trees) are given in Table 5. As can be seen, the number of trees to be sampled may be very great if a high variation is expected or if a high degree of accuracy is demanded.

In actual practice, estimates of mite density have been based on weekly sampling of 40-50 trees (5 leaves/tree) per 20 acres of orchard. The results have been satisfactory in the case of the yellow and twospotted species, but have been less satisfying with the European red mite. Further studies on spider mite sampling methodology are warranted to provide growers and pest control consultants with a confidence in predicted results now lacking in many sampling schemes.

Control Tactics

Biological control. In the absence of chemicals applied during the summer for control of other pear pests, spider mites seldom reach damaging levels. Studies in an unsprayed orchard showed that over an eight-year period (1964-71) spider mite populations never exceeded a monthly average of 0.04 mites per leaf and did not cause significant leaf injury during this period (Westigard 1973b). In commercial orchards which discontinue standard pesticide programs, spider mites are usually brought under biological control within a period of 1-2 years (Fig. 3).

Although there are several known predators of spider mites on pear in southern Oregon, the predaceous mite *M. occidentalis,* is thought primarily responsible for providing economic control. Despite the high potential for achieving biological control of the spider mite species there are several complicating factors in the actualization of this program. First, the control of other pest species requires insecticidal treatments during the summer for their control. These chemicals have generally been destructive to predator mites. This has especially been true since the appearance of resistance to organophosphates by the pear psylla and substitution of these materials with chemicals more toxic to the predator mites.

Table 5.
Sample size required to estimate
twospotted spider mite density on pear trees with specified precision.[a]

Confidence level	Confidence interval (± %)	No. of trees to sample, with coefficient of variation (%)		
		cv 100	cv 200	cv 300
90%	10	272	1089	2450
	20	68	272	613
	40	17	68	153
	60	8	30	68
95%	10	384	1537	3457
	20	96	384	864
	40	24	96	216
	60	11	43	96

[a]Taken from: Westigard and Calvin (1971).

A second complicating factor is the serious economic effects due to spider mite injury that occur in the transition period between the discontinuence of synthetic acaricides for control and the time required to achieve control with predaceous mites. Examination of Fig. 3 shows that during the transition period populations of spider mites exceeded 5-10 mites/leaf on several occasions. In this orchard both fruit size and fruit set were affected because of 1967 mite damage (Table 6). Also, it has been our experience that some use of acaricide may be necessary in years when predators failed for one reason or another to give adequate control of spider mites. Keep in mind that the twospotted mite was reported as causing severe injury in some years prior to the introduction of synthetic organic pesticides.

Ideally, when an acaricide is found to be needed it should (1) give control of the target mite, but (2) not cause resurgence in target mite species, and (3) be minimally disruptive to the predaceous mite, *M. occidentalis.* Tests conducted in 1969-70 rated several acaricides for these characteristics. Table 7 gives the results of these tests which indicated that several materials were available that met the above criteria. As can be seen, not only were there differences between materials, but rates of the same acaricide gave somewhat different results.

Despite the difficulties discussed above, a program of biological control of spider mites was utilized by many southern Oregon growers between 1968 and 1972. This program relied heavily upon chemical prebloom programs for control of the pear rust mite, San Jose scale and pear psylla and upon azinphosmethyl and oil for codling moth and psylla control during the summer (Westigard 1971).

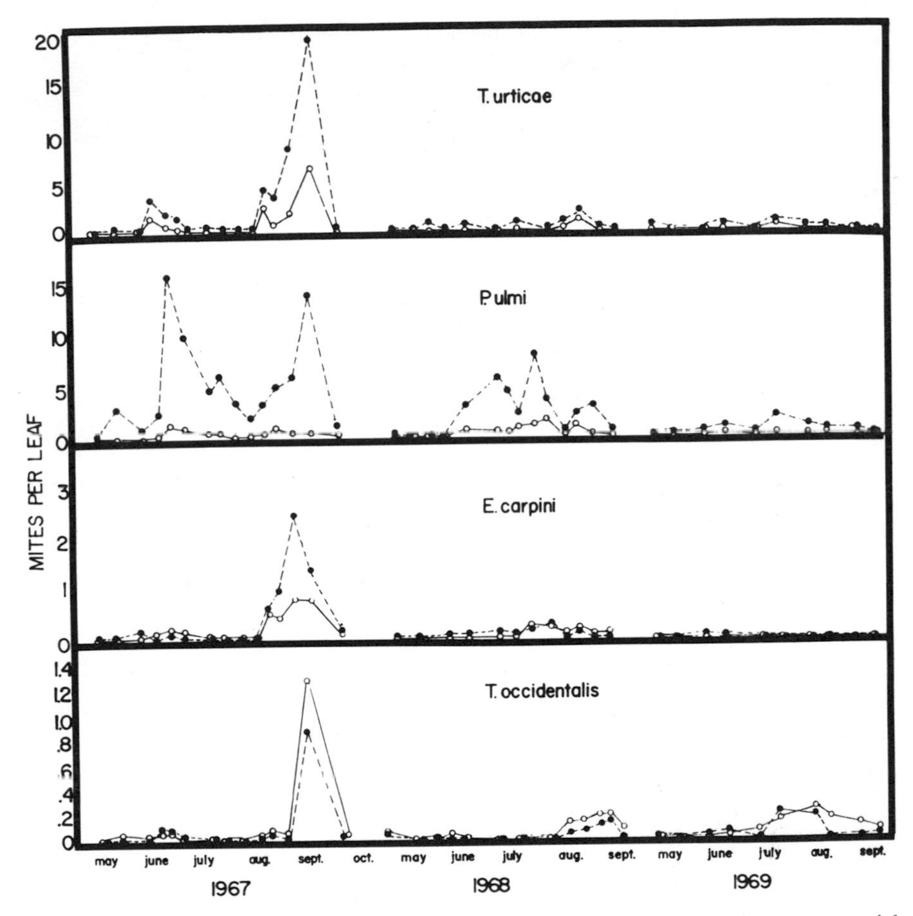

Figure 3. Population trends for spider mites and predaceous mites in a commercial orchard following discontinuance of acaricide treatments. Dashed lines indicate post-embryonic stages and solid lines egg stages. *T. occidentalis* = *Metaseiulus occidentalis.* [Taken from: Westigard (1971)].

Chemical control. The spider mites attacking pears appear, for the most part, to have been induced to pest status by the use of synthetic pesticides which were introduced following World War II. However, there is evidence to indicate that prior to this time the pear growers did suffer periodically from losses attributed to the twospotted species. Reports from the area extension horticulturalist in 1926 and 1930 indicated that the species was a major problem. In fact, the first synthetic pesticide used (DN111) was available on a commercial basis in 1942, three years before the use of DDT in the area.

Table 6.
Effect of spider mite feeding
on fruit size and fruit set of Bartlett pears.[a]

Summer pesticide program	Fruit size[b] Avg. no. fruit per carton class 1967	% fruit set 1968
Miticide	146 a	29.1 a
No miticide	156 b	15.9 b

[a]Taken from: Westigard *et al.* (1967).
[b]Means followed by the same letter are not significantly different.

Table 7.
Rating of some registered acaricides for their suitability in an integrated control program for the twospotted spider mite. Medford, 1969-70.[a]

Material and rate A.I. per 100 gal. water		Rating[b] (1 = best, 11 = worst)
Ethion	0.50	1
Plictran	0.13	2
Supreme oil	1 gal.	3
Ethion	0.13	4
Plictran	0.25	5
Phosalone	0.75	6
Dicofol	0.17	7
Acaralate	0.13	8
Acaralate	0.25	9
Phosalone	0.13	10
Dicofol	0.70	11

[a]Modified from: Westigard *et al.* (1972).
[b]Rating based on (1) degree of direct control, (2) lack of resurgence in twospotted mite density and (3) survival of predator mite *M. occidentalis.*

Subsequent to 1942 about 14 synthetic pesticides have been introduced and used for spider mite control but primarily because of resistance there is currently only one miticide available that has wide spread use (Table 8). With the exception of most of the organophosphates and chlordimeform, many of the chemicals used for mite control have been highly specific in activity and have not been observed to cause major disruption to non-target species except predaceous mites. Even with *M. occidentalis,* some acaricides have been used selectively to minimize the mortality to this species (Table 7).

The European red mite is the only spider mite species which currently receives prebloom or post harvest chemical treatment for its control. With this species the overwintering eggs are in an exposed situation and can be contacted with spray materials. Petroleum oils have been used for this application since the European red mite appeared in 1952, with no indication of a decrease in effectiveness. In general, the control with the prebloom oil sprays will be improved by delaying the treatment as long as possible into the delayed dormant or pre-pink period. Treatment at this time may also give a degree of control of the other spider mites that are beginning to move from overwintering sites beneath the bark scales to the opening fruit buds.

Despite the apparent effectiveness of the prebloom or post-harvest treatments in lowering spider mite density, these treatments have generally been discouraged in southern Oregon except when their use has been shown to be vital in the suppression of other pest species. These treatments have been excluded because they have not been shown to produce economic benefit through the elimination of one or more of the summer acaricide treatments. In other words, the prebloom of post harvest use of acaricides for the twospotted spider mite can be viewed only as preventative applications without regard to the IPM concept of economic thresholds or economic injury levels.

Foliar applications have traditionally been accepted by growers as the proper period for use of acaricides in the control of spider mites on pear. This foliar period extends from post bloom (petal fall) until close to harvest (mid-August for the early 'Bartlett' variety and late September for the winter pears). However, within this time frame growers' decisions regarding the need for acaricide treatment is limited by other, non-entomological, factors such as irrigation schedules, thinning, etc. In practice, growers make every attempt to include acaricide treatment along with sprays to control other pests such as the codling moth. These summer treatments are applied about one month apart and begun in mid-May. The deletion of an acaricide from one or more of these summer "cover sprays" therefore demands that it be certain that the spider mite levels will not reach economically damaging densities for 3-4 weeks or until ground conditions again allow passage of spray equipment.

Cultural control. There is some evidence to indicate that the vegetation on the orchard floor harbor populations of the twospotted spider mite and that these are capable of moving into the tree canopy (Westigard *et al.* 1967). Before recommending a program of clean cultivation however, it should be remembered that not all orchard floor plant species will support pear feeding spider mites and

Table 8.
Chronology of use pattern for synthetic acaricides used during foliar period on pears in southern Oregon.

Material	Year first used	Year discontinued	Reason discontinued
DN111	1942-43	1946	Phytotoxic
EPN	1946	1954	Lack of control
TEPP	1947	1969	Loss of registration
Aramite	1947	1960	Loss of registration
Parathion	1948	1953	Lack of control
Ovatran	1949	1955	Lack of control
Chlorobenzilate	1952	1968	Lack of control
Kelthane	1955	1972	Lack of control
Tedion	1956	1962	Lack of control
Trithion	1955	1968	Lack of control
Morocide	1968	1972	Phytotoxic
Chlordimeform	1970	1976	Withdrawn by manufacturer
Ethion	1963	1970	Lack of control
Plictran	1972	In use	–
Vendex	1976	In use	–

that spider mite hosts may also be important reservoirs for predaceous mite and other beneficial species. A more thorough study of this relationship is necessary before any definite conclusion as to cover crop management can be reached.

Significant reductions in spider mite levels following overtree irrigation has been reported from Northwest fruit growing areas. In southern Oregon pear orchards, no dramatic population reductions have been measured. There are some observational data available, however, that indicate that while spider mite density is not affected, the damage caused by these levels may be reduced by minimizing the effect of water loss associated with mite feeding.

PEAR PSYLLA

The pear psylla, *Psylla pyricola* Foerster was first reported in the United States from Connecticut in 1832. It was detected in Washington state in 1939 and in Hood River, Oregon in 1949 and in southern Oregon's Rogue River Valley in 1950. Though the initial appearance of this pest was greeted with apprehension by local growers and those responsible for pest control in the area, the destructive potential of the species was not immediately realized. However, in 1957, pear trees in the area began to show symptoms of poor growth, lower tree vigor and outright collapse. Though for a 4 or 5 year period this malady referred to as pear decline was not related to pear psylla, it has since been defined as being induced by a mycoplasma-like organism carried by this pest. Between 1957 and 1963, it has been estimated that nearly 2,000 acres of pear in southern Oregon were either killed outright or their production capabilities severely crippled.

Though the period of pear decline ravages has subsided, the pear psylla still remains an important pest of pear and currently is the most expensive pest to control in commercial orchards.

Life History

Developmental stages. The pear psylla overwinters in the adult stage which is somewhat larger and darker than the summer adult. Both males and females overwinter and mating does not occur prior to egg laying in January or early February. Most of the first eggs laid by overwintering females are deposited on fruit spurs but as the fruit buds open, eggs will be laid on exposed leaf or fruit tissue.

The pear psylla passes through five nymphal instars prior to reaching the adult stage. All nymphal stages feed on green leaf tissue and as they feed secrete a sticky "honeydew" liquid around their bodies. The 5th instar is referred to as the "hardshell" stage and is dark brown to black in color and bears prominent wing pads. This stage is active and often is found at the base of leaf petioles. In southern Oregon, there appears to be four complete generations annually (Westigard and Zwick 1972).

Dispersion. As mentioned above there are morphological differences between the overwintering and summer adult forms. The larger, overwintering adults exhibit a greater tendency for dispersal and are capable of moving substantial distances. It is typical in southern Oregon for this form to disperse from the pear orchards in early fall into surrounding vegetation. In late winter there is a reversal in this behavior and movement back into pear occurs. It is not known what portion of the population disperses and what percent remains in the orchard to overwinter.

Both the fall and winter dispersal patterns are important to the management of the pear psylla. For example, in late winter movements back into pear orchards begin in early January but may continue through April or later. Chemical treatments against overwintering adults which are normally applied in early February may be highly effective against psylla present in the orchard but normally will not have the persistance necessary to control late arriving immigrants. Thus, multiple prebloom pesticide treatments are sometimes used to achieve economic reduction.

Host plant conditions. The amount of young succulent pear foliage available appears to be an important factor in the determination of pear psylla densities. The actively growing points on the pear tree seem to be preferred for oviposition. This is especially evident in the late winter when the vast majority of eggs from overwintering females are found around swelling buds. Also, during the summer months eggs are most commonly laid on the terminal leaves of actively growing pear shoots. The sampling methods developed have utilized this relationship to measure pear psylla egg and nymph densities through the year (Burts and Retan 1973).

Host range. Although the adult pear psylla and occasionally the egg stage can be found on other hosts, the insect can complete its development only on pear. Even within the genus *Pyrus,* there are species which appear to be unable to support the completion of psylla development. This monophagous habit of the pear psylla offers some promise in its management.

Damage

The pear psylla is known to cause at least three types of damage to pear. Each of these damage types is associated with different psylla densities and thus have different economic injury levels. In addition, each damage type has control tactics which are most suited to its prevention.

Pear decline. As previously described the cause of this disease has been attributed to a mycoplasma-like organism of which the pest psylla is the vector. The symptoms vary but may be expressed by lowered tree vigor, poor fruit set, small sized fruit or by death of the tree. The symptoms are caused by sieve-tube necrosis at or below the graft union and their severity is related to the origin of the rootstock material (Batjer and Schneider 1960). Cultivars grafted onto *Pyrus pyrifolia* or *P. ussuriensis* are more susceptible than those on *P. communis.* Since the rootstocks themselves are usually propagated from seeds rather than clonal root cuttings there is a natural variation in pear decline susceptibility even within the resistant or susceptible understocks. With disease carrying insects the pest density necessary to cause economic damage is generally very low and in the case of pear psylla-pear decline relationship, it was presumed to be near zero. In this situation the most appropriate control tactic was the introduction of resistant rootstock material. Horticultural research demonstrated that there were several rootstocks which were not affected by the mycoplasma organism and these have been used in many of the new pear plantings in the southern Oregon area (Westwood *et al.* 1971).

Psylla toxin. A second type of injury attributable to pear psylla is that caused by the injection of a toxicogenic substance by nymphal stages. Symptoms of this disorder are similar to those of slow decline including poor tree vigor and lowered productivity. However, the effects of the toxin are generally related to high psylla densities over several years and are found on cultivars grafted onto both decline-tolerant as well as susceptable rootstocks. Damage from psylla toxin has not been observed in southern Oregon under conditions of low to moderate psylla infestation, and it is believed that the economic injury level for this damage type is only reached during years when psylla numbers go relatively unchecked by chemical or biological controls.

Psylla honeydew. In the process of feeding, psylla nymphs secrete a sticky substance called honeydew. This material may drip onto fruit causing a dark russeting on the surface. If the marking is excessive, downgrading of the fruit will occur and crop value will be reduced. The presence of copious amounts of honeydew at harvest have resulted in picker complaints and increased harvesting costs.

In contrast to pear decline and the effects of psylla toxin, injury from psylla honeydew represents direct fruit damage and is of the utmost concern to growers. The current spray costs of over $100/acre/year for psylla control are directed at avoiding this type of damage. Unfortunately, we have conducted no critical studies to relate psylla density to potential fruit damage and without this information the implementation of IPM for this pest has been curtailed. Some work has been conducted on economic injury levels in Washington state, but has not as yet been substantiated under southern Oregon conditions (Brunner 1975).

Sampling

The sampling methodology for pear psylla has been developed to measure densities of various stages of this pest at various times of the year. While these sampling procedures have not been related to a known economic injury level, they can be used in a generalized sense based on experience to indicate the presence of potential damaging populations.

Adults. Sampling adult pear psylla is carried out be jarring limbs with a rubber covered piece of wood dowling and recording the number of adults falling onto a 18" x 18" cloth covered catching frame. This procedure is followed from the prebloom period through the summer months. The information is used to determine the (1) degree of control obtained following chemical treatment, and (2) population structure which, in conjunction with counts of immature stages, is used to determine control strategy. Limb jarring also yields data on the density of several important psylla predators and parasites.

Immatures. As mentioned previously the pear psylla adults deposit eggs on actively growing points of the pear tree. Thus, the sampling unit changes through time. Table 9 gives the sample units used during these seasonal changes.

Table 9.
Sampling units for
evaluation of immature psylla density.

Period	Psylla stage	Unit
Dormant to bud swelling (January to March)	eggs	unopened fruit spurs
Prepink to petal fall (March 15 to May)	eggs & nymphs	leaves and fruit from fruit spurs
Summer (May to September)	eggs & nymphs	terminal and basal leaves on growing pear shoots

Control Tactics

Biological control. In southern Oregon the role of predators appears to play a significant role in the natural control of the pear psylla at least under certain conditions. In an 8-year study (1964-1971) conducted in an unsprayed 'Bartlett' pear orchard, populations of the pest did not reach levels to cause commercial downgrading of fruit due to psylla honeydew (Westigard 1973b).

There are over 30 predators and parasites which have been reported from North America as feeding on the pear psylla (McMullen and Jong 1967). In southern Oregon the most important appear to be the common green lacewing, *Chrysopa carnea* Stephens, and the mirid, *Deraeocoris brevis piceatus* Knight. The latter was shown to consume an average of about 400 psylla eggs and nymphs during their development period (Westigard 1973a).

In unsprayed orchards the predator complex generally begins to appear in mid to late May and their numbers peak in late June or July. Typically pear psylla densities will be economically suppressed during this period (Fig. 4). Cool temperatures in early summer appear to delay the influx and buildup of predaceous species and under these conditions commercial psylla suppression may not be achieved.

As mentioned in the section on codling moth, most organophosphate pesticides used for codling moth control will substantially reduce the density of psylla predators. An example of the effects of these pesticides on *D. b. piceatus* is given in Table 10. As seen, the 24-hour mortality to the nymphal stage of this mirid ranged from 89-100% when exposed to direct chemical application. The use of *D. b. piceatus* for pear psylla control will depend upon finding non-disruptive control tactics for codling moth.

Chemical control. When the pear psylla first made its appearance in the Rogue Valley in 1950 the organophosphate chemicals already being used in the area during the summer period were also effective against it. About 1961 the

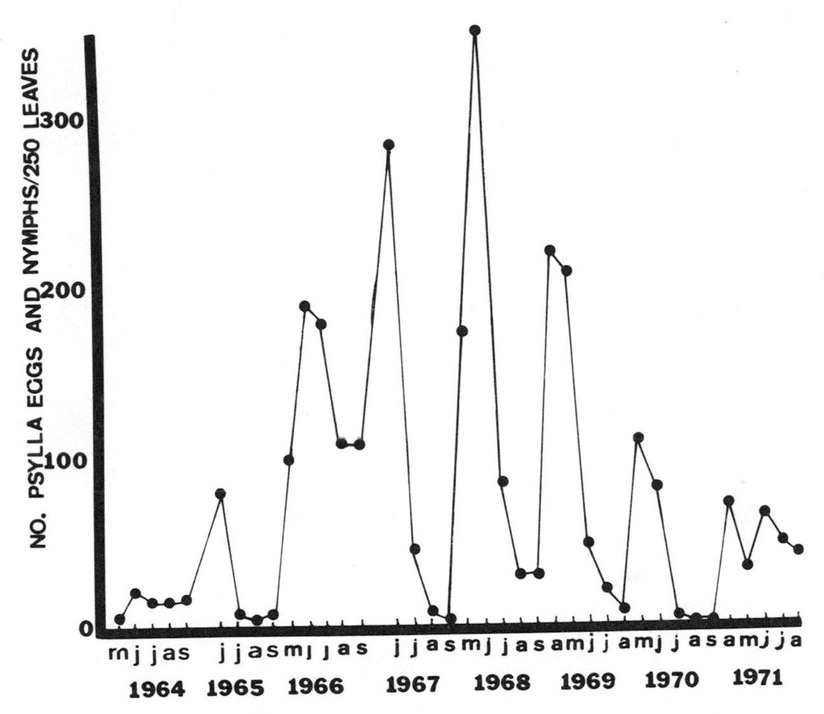

Figure 4. Population trends of the pear psylla in an untreated Bartlett pear orchard. 1964-71. Suppression of psylla population by parasitoids usually occurs during July-August. [Taken from: Westigard *et al.* (1968)].

registered organophosphates were no longer effective and growers substituted chloronated hydrocarbons such as dieldrin. These compounds lasted only about four years until resistance made them non-commercial. In 1962 azinphosmethyl, an organophosphate effective against psylla, was introduced and gave commercial control until 1965 when it too was lost due to resistance. At present, amatraz is the only effective chemical available for summer use.

In 1965, a dormant spray treatment directed exclusively at overwintering pear psylla adults prior to oviposition was first used by southern Oregon growers. This treatment, developed by Burts (1968) in Washington state was introduced to compensate for the absence of effective summer pesticides which had been lost to resistance. Perthane®, the chemical initially used in 1965 at this timing still remains effective. However, because Perthane® exhibits relatively short residual activity the degree of control obtained depends upon its being used at the appropriate time. Two techniques have been used to determine the appropriate timing for this dormant treatment. These include the disection of

Table 10.
Effects of direct application
of pesticides on *D. brevis piceatus* nymphs.[a]

Material	Lb. a.i. per 100 gal. water	Number treated	% mortality 24 h
Azinphosmethyl	0.13	15	100
	0.50	17	100
Phosmet	0.50	18	100
	1.00	18	100
Phosalone	0.37	18	89
	0.75	17	94
Untreated	–	25	0

[a]Taken from: Westigard (1973).

Table 11.
Egg deposition by overwintering psylla
in relation to accumulated degree days over 43°F from January 1.[a]

Year	Date of first psylla egg	Date, no. degree days over 43°F = 250
1970	January 27	January 25
1969	February 14	February 26
1968	February 10	February 7
1967	February 8	February 15
1966	February 10	February 19
1964	February 14	February 18
1963	February 11	February 8
1960	January 24	February 6
1959	February 5	February 1
1958	February 9	February 7
1957	February 18	February 18

[a]Taken from: Westigard and Zwick (1972).

overwintering female psylla to examine them for the presence of mature eggs and the use of temperature accumulation to predict egg development. This latter method, suggested by Burts (1968) in Washington state, accumulates maximum temperatures over a base of 43°F beginning January 1. In southern Oregon it has been found that egg laying begins when these totals reach about 200-250 degree days (Table 11). Recent developments in host plant masking and the probable registration of more persistant adulticides for prebloom treatment may cause a relaxation in the critical timing now required to obtain commercial control with the dormant treatment.

In recent years there has been an increased use of pink bud timing to obtain additional reduction in psylla numbers. As was the case for the dormant spray, this timing was necessitated by the appearance of resistance to organophosphates used during the summer months. The advantage to the pink bud spray is that pear psylla populations are predominately in the early nymphal instars at this time and are more susceptible to the insecticides.

The history of summer use patterns of insecticides for psylla control has been discussed above. These chemicals are generally added to the codling moth sprays which are applied 3-4 times per season. With less effective compounds being available for psylla control, the summer sprays are timed for the period when the majority of the pear psylla are in the younger, more susceptible nymphal instars.

Host plant resistance and host plant modification. The pear psylla is a monophagous insect and can complete its development only on pear. The genus *Pyrus* contains 20-25 species of pear of which three are commonly cultivated for their fruit and several others used for understocks. The common European pear, *P. communis,* is of ancient origin from which several hundred varieties have been derived. Studies during 1969-1970 in southern Oregon revealed that the pear psylla did not prefer to lay eggs or would not complete their development on several *Pyrus* species. For the most part, the commercially grown varieties of *P. communis* were highly preferred while those species from Asia except for *P. pyrifolia* were the least suitable hosts. Unfortunately, no significant degree of host resistance was found in *Pyrus* cultivars which bear fruit of commercially acceptable quality (Westigard *et al.* 1970).

The pear psylla seemingly prefers young succulent foliage for ovipositional sites and sampling techniques for this pest are based on selecting actively growing buds or shoots as the sample unit. A chemically induced reduction in overall shoot growth has been found to cause a corresponding reduction in pear psylla numbers. Using the tree growth regulator Alar® at 2000 ppm applied in May and June, overall population levels of pear psylla eggs and nymphs were reduced by 50% and honeydew damage to fruit by 75%. Although these studies are not yet completed, it seems that the use of tree growth controlling chemicals may be an important control tactic in regulation of pear psylla.

As described above the pear psylla is highly selective in its choice of host and in its selection of suitable oviposition sites. This would infer that the species must receive the correct chemical clues or stimuli from its host before egg laying is triggered. Zwick and Westigard (1978) found in studies conducted in the Rogue Valley and elsewhere that treatments with petroleum oils in the dormant period prevent egg laying by the pear psylla for up to six weeks (Fig. 5). We believe that these treatments may be acting to inhibit the production or reception of chemical stimuli produced by the pear host which is necessary to evoke oviposition. This use of oil sprays as ovipositional deterants have been useful in the management of pear psylla. The treatments, applied in early to mid January, allow growers greater flexibility in the timing of the dormant adulticide treatment by extending the pre-ovipositional period and allowing for a more complete return of adult psylla from sources outside the orchard.

Cultural control. Cultural practices that increase the vigor or lengthen the period of new shoot growth will increase the severity of pear psylla attack. These practices include pruning, fertilization, and irrigation. Careful management of

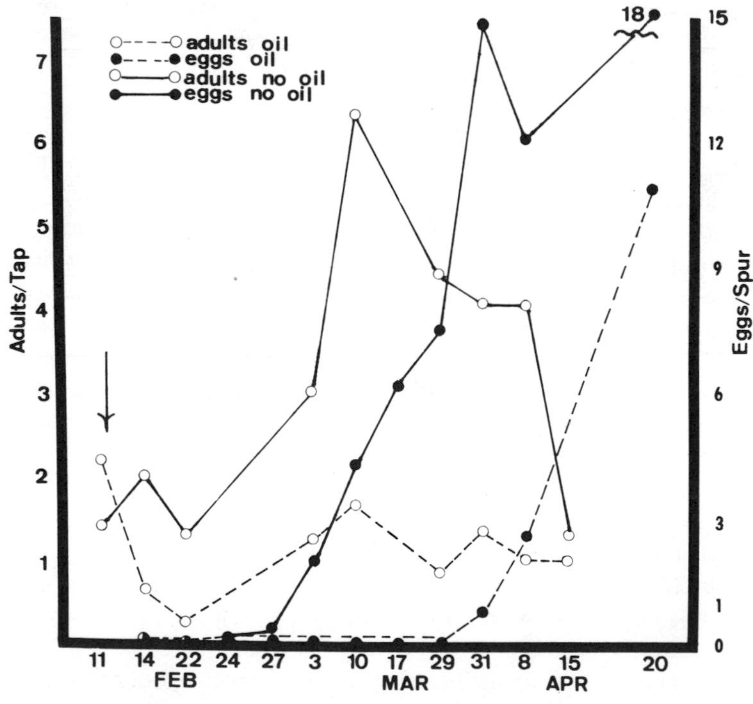

Figure 5. Effect of oil sprays on delay of egg laying by the pear psylla. Arrow indicates date of treatment. [Taken from: Zwick and Westigard (1978)].

Table 12.
Effect of overtree irrigation
on pear psylla density and honeydew damage to Bartlett fruit.

Sprinkler irrigation	% change in psylla 24 hrs after irrigation	% honeydew damaged fruit at harvest[a]
Overtree	+ 16	4.2
Undertree	+ 18	14.7

[a]% damage following 6 biweekly irrigations during summer of *ca.* 3" water at each irrigation.

these horticultural aspects to avoid excess tree stimulation should be practiced. One cultural practice, overtree irrigation, can be used advantageously to reduce psylla injury by washing potentially damaging honeydew from trees. It has been found that this method of irrigation can reduce the incidence of honeydew injured fruit by 50-60%; however, the practice did not result in substantially lowered psylla densities (Table 12).

SAN JOSE SCALE

The San Jose scale, *Quadraspidiotus pernicisus* (Comstock), along with the codling moth was one of the earliest recognized pests of pear in southern Oregon (Cordy 1977). As early as 1890 growers were concerned about the tree and fruit losses attributable to this scale. While the San Jose scale is currently controlled to the point where little tree damage results, it still causes severe fruit downgrading in many Rogue Valley orchards.

Life History

In southern Oregon, as in most temperate climates, the San Jose scale overwinters in the 1st nymphal instar referred to as the 'blackcap' stage. Both males and females overwinter. Growth is resumed in late winter following diapause termination and the occurence of warmer temperature regimes. Immature male scale will pass through two molts into a pupal stage from which the fragile 2-winged adult male emerges. Females pass through three molts before reaching maturity and after mating give birth to live offspring called crawlers. These young (except for the mature males) are the only motile forms exhibited by this insect. Generally crawlers move but a short distance from the mother and settle on woody tissue or on leaves and fruit (Gentile and Summers 1958). In southern Oregon there are two complete generations of San Jose scale per year. The first generation crawlers are usually found in early June and continue to appear through late July. Later appearing crawlers of the 1st generation overlap slightly with the crawlers of the second generation. Crawlers of the last generation appear in August and usually can be found until fall frosts occur.

Damage

Under conditions of relatively high population densities the San Jose scale may cause limb dieback, reduced tree vigor, or even the outright death of the pear tree. However, even in untreated orchards, attacks of this severity are infrequent. Another type of damage, that caused by crawler settlement and feeding on the pear fruit is very common in southern Oregon, and because of the strict grades and standards set on fruit quality this type of injury often results in severe crop losses. The presence of one live scale or the feeding mark caused by one scale will prohibit the fresh shipment of pears to most European markets, and over two scale per fruit will drop the grade from the U.S. 1 to the Fancy category. The cost of such grade changes are discussed in the section dealing with the economics of pear pest management.

Scale density and damage. The allowable scale damage that would offset the cost of current control practices depends upon the value of the various pear varieties. Using the 'Bosc' variety as an example and pesticide costs of approximately $30/acre, the allowable fruit loss to San Jose scale in the absence of any costs for control would be in the range of a 2% fruit infestation. The scale density necessary to cause this 2% loss would be an economic injury level. Using scale infestations of fruit spur wood (see section on sampling) as an indicator of the economic threshold for fruit damage (economic injury), it was estimated that an infestation of fruit spurs exceeding 1-4% will result in economic losses at harvest (Fig. 6).

Sampling

Several sampling methods have been used for San Jose scale including the taking of twig and shoot, bark or fruit samples. In general these have been used to evaluate the efficiency of various control tactics, especially pesticides. In the practice of integrated pest management it is helpful to select a sample unit which has value in forecasting potential economic loss. In southern Oregon the pear fruit is considered to be the plant part most sensitive to injury, and the sample unit most predictive of losses to the fruit is believed to be the woody portion of fruit spurs. The choice of this unit was based on the low vagility of crawlers and the fact that this unit with its attached leaves and fruit could also be used in the density determination of other pear pests. Samples taken from the upper ½ of the tree generally support higher scale numbers and are better predictive units of damage potential.

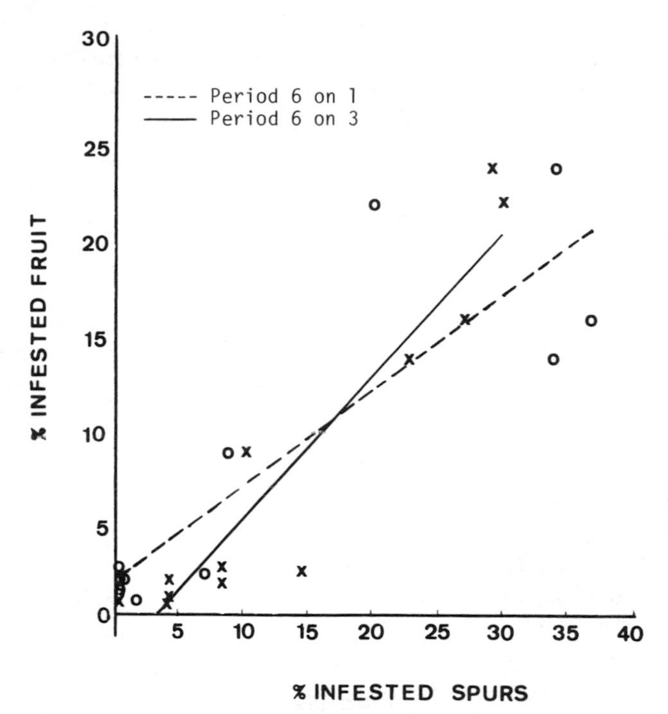

Figure 6. Relationship between the percent infested spurs in period 1 (April) and 3 (July) to percent infested fruit in period 6 (September) following delayed dormant application of pesticides for control of San Jose scale. Data points (X and 0) represent the average percent infested spurs and infested fruit from 4 single-tree replications for each of 10 chemical treatments and an untreated check. See Table 13 for designation of periods. [Taken from: Westigard and Calvin (1977)].

In the case of San Jose scale, it was found that the total number of scale on fruit spurs need not be counted, but rather the percent infested spurs was a better predictor of percent infested fruit at harvest (Table 13). In addition samples taken in April, June or early August were more reliable than those taken in May or late August.

Control Tactics

Biological control. Data from 1964-1971 from unsprayed 'Bartlett' orchards showed the San Jose scale infestation of fruit at harvest ranged from 1% in 1964 to 92% in 1967 (Westigard 1973b). Based on tolerable injury levels for the various varieties the infestation level recorded surpassed tolerable economic levels in 5 of the 7-year study period. Thus, it seems that the native parasite and predator complex cannot be relied upon to give economic pest suppression in

Table 13.

Correlation coefficients between % San Jose scale infested Bosc pears in period 3 or period 6 and total number of scale or % infested spurs found in previous sample periods, Medford, Oregon. 1973.[a]

Period infestation sampled[b]	Period fruit sampled	Regression coefficient	
		Total no. scale	% infested spurs
1	6	0.567**	0.708**
2	6	0.341**	0.591**
3	6	0.554**	0.803**
4	6	0.497**	0.838**
5	6	0.661**	0.694**
1	3	0.541**	0.634**
2	3	0.307*	0.531**

*Significantly different from zero at P = 0.05.

**Significantly different from zero at P = 0.01.

[a]Taken from: Westigard and Calvin (1977).

[b]Period 1 - late April; 2 - late May; 3 - late June - early July; 4 - early August; 5 - late August; and 6 - mid - late September.

Table 14.

Effects of laboratory pesticide treatments on San Jose scale crawlers.[a]

Material and lbs. a.i./100 gal.		% crawlers dead but settled 14 days after treatment	% mortality of crawlers at days after treatment	
			14 days	28 days
Parathion	0.50	78	94	22
Phosalone	1.00	52	99	68
Ethion	0.25	92	32	0
Ethion	0.50	—	82	58
Diazinon	0.40	80	—	—

[a]Modified from: Westigard (1977).

regard to tolerable fruit injury levels. On the other hand, there are several natural agents that attack the San Jose scale and these, if left undisturbed by pesticide treatments, appear to be able to hold scale levels below the level causing severe loss in tree vigor. These natural enemies include several coccinillid species, a predaceous mite, and the parasite, *Prospaltella perniciosi* (Tower). This latter species was found in 1970 to have attacked over 80% of the scale on untreated 'Bartlett' trees (Westigard 1970).

For the San Jose scale, as well as several other pear pests, the potential for biological control would be quite promising except for the high standards of fruit quality demanded either by law or by the general public.

Chemical control. Because the San Jose scale overwinters as a young nymph primarily on exposed sites on pear limbs or shoots, a chemical treatment applied in the dormant period usually gives commercially acceptable control. This treatment timing has been used effectively by Rogue Valley growers since the early 1920's. Currently the spray, usually containing oil plus either sulphur or an ogranophosphate insecticide, is applied prior to bud separation because applications of the compounds at a later date can be injurious to pear buds. Tests in 1974 indicated that the oils used alone were usually sufficient to obtain control but combination sprays may add to scale mortality as well as be effective against other pest species (Westigard 1977).

The dormant or the delayed dormant treatment will generally provide better control than post bloom treatments. This is probably due to the improved tree coverage offered at this time attributed to the absence of foliage. In addition, the prebloom timing is less disruptive to non-target beneficial species than the sprays used during the foliar period.

The summer treatment for San Jose scale control is usually timed to coincide with the appearance of the crawler stage. As previously discussed, the crawlers of the first generation usually appear in early June and those of the second generation in mid August. Treatments during these periods, while they reduce population numbers, cannot be relied upon to give commercial control in the absence of prebloom sprays. In addition to the poorer spray coverage obtained during the summer, there are two other reasons why the sprays directed at crawlers do not result in commercial scale reduction. First, the emergence period spans about two months and the residual activity of the registered pesticides is not nearly of this duration. Tests in 1969 showed that while some materials killed crawlers for a 14-day period their effectiveness was dissipated at least by the end of a 4-week period (Table 14). While summer treatments timed at 14-day intervals may be effective they are rarely used in southern Oregon because of the growers preoccupation with other necessary horticultural practices.

Another possible timing during the foliar period is that directed at males just prior to adult emergence (Downing and Logan 1977). Since these forms appear over a shorter period and because of their greater vagility, it was presumed that they would be more susceptible than the crawler stage to summer chemical suppression. This timing did reduce scale infestation over that

Table 15.
Control of San Jose scale
obtained by treatments directed at male or crawler stages.

Timing[a] and date	% infested fruit
1st male - April 19	18.0
1st male + 7 days - April 27	35.0
1st male + 14 days - May 6	25.5
1st crawler + 7 days - June 13	35.5
Untreated	71.3

[a]Diazinon at 0.5 lbs. a.i./100 were used in these studies.

produced by sprays directed at the crawlers; however, the data also demonstrates the poor control of San Jose scale which can be expected from treatments restricted to the foliar period (Table 15).

PEAR RUST MITE

The pear rust mite, *Epitremerus pyri* (Nalepa), was first identified from southern Oregon pear orchards in 1931. Between that time and the present there have been reports of sporatic but severe damage due to this eriophyid mite. While the species attacks both pear fruit and foliage, it is considered a direct pest as leaf injury is seldom of economic importance.

Life History

Little is known regarding the details of the biology of the pear rust mite, but information is available regarding its seasonal intratree distribution. This aspect of the life history is of importance to IPM because it affects sampling procedures. The pest overwinters in bark crevices or behind loose bud scales, predominately on 2 to 3-year old wood (Fig. 7). Emergence from these overwintering sites coincides with the separation of fruit buds. These developing clusters become the initial feeding sites for the rust mite. A period of time is spent within the fruit cluster when damage to fruit may occur. As the clusters open and leaves expand the rust mite moves to leaf tissue, though a residual population may be found on the fruit through most of the summer. During this period the rust mite passes through several generations but the exact number is unknown. Also uninvestigated, is the fecundity and longevity of the mite. Observations would indicate that there are several generations per year and that these may be favorably influenced by the availability of young succulent tissues.

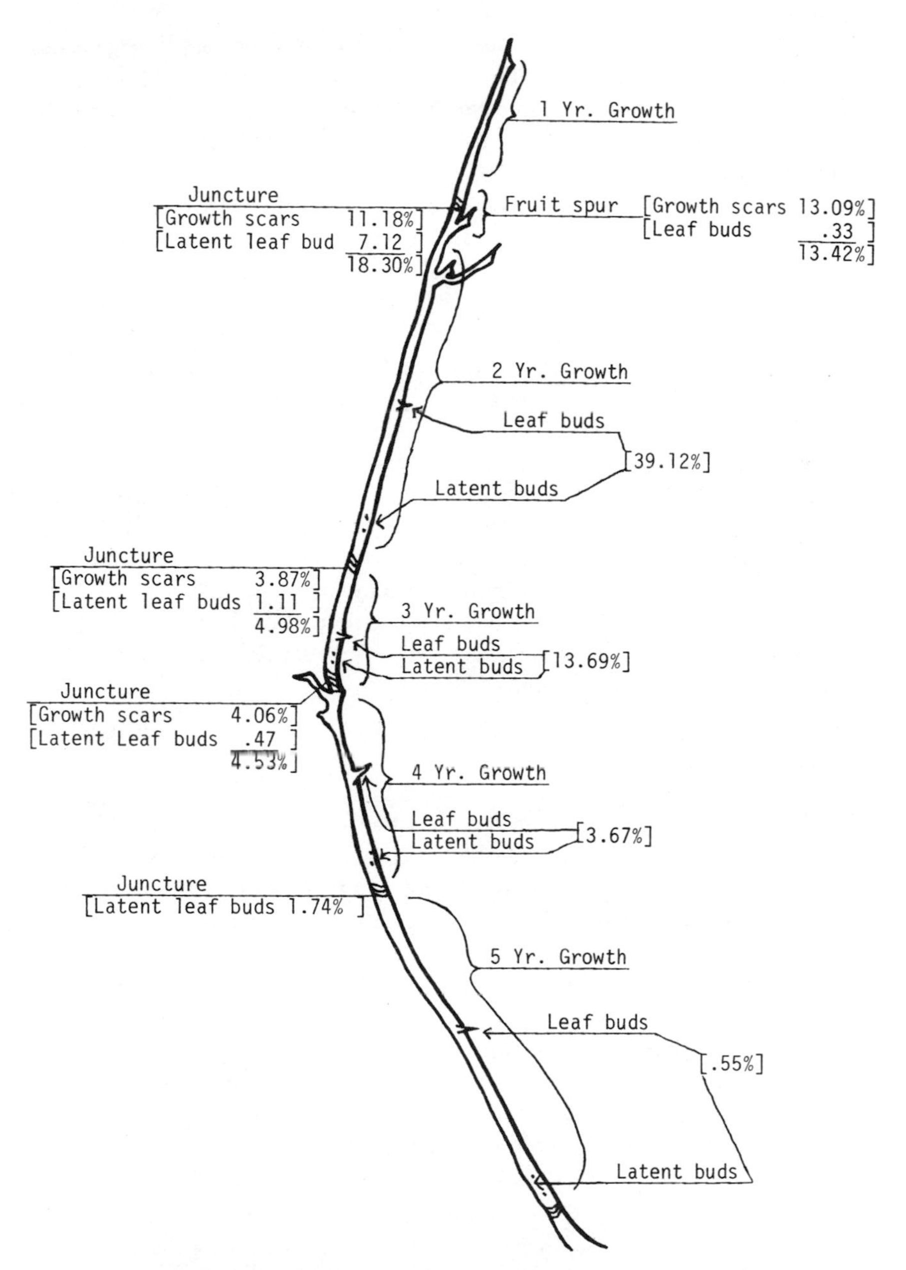

Figure 7. Overwintering distribution of the pear rust mite on 5-year wood sections of pear limbs, December, 1972. [Taken from: Westigard (1975)].

Damage

The pear rust mite feeds on the surface of fruit and foliage causing a bronzing of the tissue. 'Bartlett' fruit injured by the rust mite is usually acceptable for cannery use but for fresh market a heavy russett covering over 5-10% of the surface will result in downgrading or perhaps culling. Other clear skinned pear varieties such as 'Comice' or 'D'Anjou' which are sold almost exclusively on the fresh markets are particularly susceptible to rust mite damage. The normally russetted 'Bosc' variety is seldom downgraded due to rust mite feeding.

Mite density and damage. In 1974, the relationship between seasonal rust mite density and russett damage was evaluated. For these tests an acceptable level of injury was based on current control costs and the type and degree of rust mite damage. Based on a spray cost of $10/acre/year and a return to the grower of about $1000 per acre for fruit value, a loss over 1% of the fruit would exceed the cost of control. The population density responsible for a loss of this magnitude would then be an economic injury level. The results obtained indicated that pear rust mite densities exceeding an average of 5 per fruit on one or more sample dates resulted in excess of 1% fruit damage, *i.e.* fruit with over 5% of the surface russetted. In these studies there were indications that the economic injury level may be higher in the early part of the season (May) than in June or July (Fig. 8). However, economic injury levels would be lower on more valuable varieties such as 'Comice'.

Sampling

The pear rust mite may be sampled to detect its presence, to evaluate control tactics or to determine the potential for economic injury. The first two sampling objectives are probably best accomplished by fruit buds (early season) or leaves (summer) serving as the sample units. However, as suggested by the dispersion pattern exhibited by the rust mite, these sample units may not be indicative of potential fruit injury. In the studies on rust mite injury cited above, the sample selected was the pear fruit, as this represented the basic economic unit influenced by mite feeding. In addition, fruit samples in the upper portion of the pear tree consistantly supported higher rust mite densities (Fig. 9) and therefore were used in the sampling system developed. This system, using 5 fruit from the upper ½ of each tree, was used in estimating the sample size (#of trees) required to obtain a reliable estimate of mean population density.

Based on density data taken from untreated 'Bartlett' trees, the required sample size was calculated for various sample periods (Table 16). As can be seen the sample size decreased with time indicating decreasing variation between sample means. From a cost standpoint, using the confidence interval of ±10%

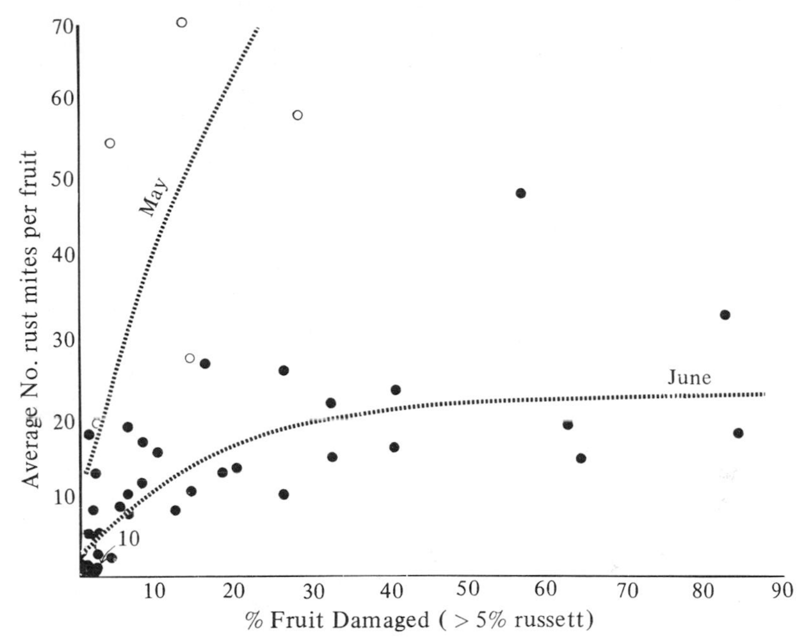

Figure 8. Relationship between rust mite density in May and June and fruit injury at harvest (August) on Bartlett pears, 1974. [Taken from: Westigard (1975)].

would result in a sample size so large as to be impractical. In most cases it would seem that a confidence interval of $\pm 30\%$ or even higher would have to be employed.

Control Tactics

Biological control. In pear orchards which do not receive pesticide treatments, the pear rust mite, like the codling moth, is a persistantly destructive pest. Between 1965-1971, nearly 50% of unsprayed 'Bartlett' fruit was downgraded or culled because of rust mite damage (Westigard 1973b). Thus, the potential for biological control to achieve commercially acceptable levels of rust mite suppression with the native biological control complex in southern Oregon appears poor.

Chemical control. The pear rust mite has historically been controlled in the prebloom or post harvest period. Treatments primarily directed toward the San Jose scale generally have been effective on rust mite, also (Westigard and Berry 1964). Occassionally, because of wet ground conditions, the prebloom treatments are not applied and severe rust mite damage may occur. In 1968 a study was conducted to evaluate timing of application in relationship to control and fruit injury. These studies showed that prebloom treatments were necessary in

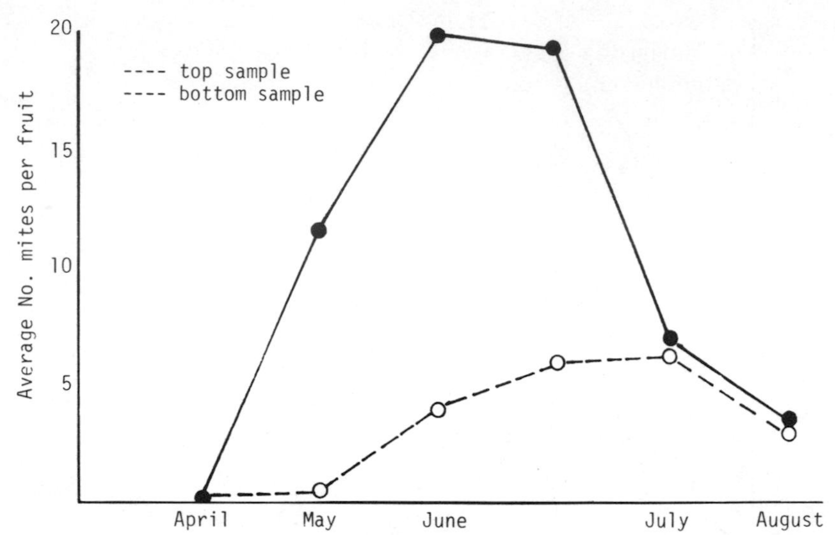

Figure 9. Rust mite density on fruit from top and lower areas of untreated 'Bartlett' pear trees, 1973. [Taken from: Westigard (1975)].

Table 16.
Sample size required
to estimate pear rust mite density.[a]

	Confidence level 95%			
	No. of trees to sample with coefficient of variation (%)			
	May	June 6	June 20	July
Confidence interval (± %)	CV 185%	CV 107%	CV 82%	CV 69%
10	1314	440	258	183
30	146	49	29	20

[a]Taken from: Westigard (1975).

order to minimize damage. The later the rust mite control was applied the more severe the fruit russett (Table 17). Currently there are many pesticides available for rust mite suppression but the choice of a particular compound usually depends upon it being effective on other pear pests in addition to pear rust mite.

Table 17.
Timing of pesticide application for control of *E. pyri* in relationship to injury to Bartlett fruit. Medford, Oregon. 1966.[a]

Time of application[c]	No. *E. pyri*					Fruit injury[b]	
	Apr. 22[d]	May 6[e]	May 24	June 8	June 27	% fruit surface marked	% fruit with russett
Delayed dormant (Mar. 12)	9	0.4	1.2	0.3	0.5	5.0	2.0
Pink bud (Mar. 31)	3	0.9	1.3	0.5	0.5	5.8	6.0
1st cover (May 15)	409	15.1	0.6	0.5	0.8	9.3	64.0
2nd cover (June 15)	559	20.3	48.3	14.1	0.0	21.0	94.0
Untreated	1234	16.1	18.6	10.4	6.8	29.5	86.0

[a]Taken from: Westigard (1969).
[b]At harvest July 29.
[c]Treatment of 1 lb. endosulfan applied to four 4-tree replicates.
[d]Total *E. pyri*/25 leaf punches, 3/4 in. diam.
[e]Avg. no. *E. pyri*/leaf; sample size of 80 leaves.

SOME ECONOMIC AND HORTICULTURAL ASPECTS OF IPM

Economic Aspects

One of the important aspects of integrated pest management deals with crop economics interfaced with pest control decisions. The IPM system requires that pest density be related to damage potential and that damage be shown to have real economic impact.

Components of crop value. There are two facets of pear culture economics that bear on IPM. First, there exists a potential crop value aside from pest damage considerations and second, the interplay of pest levels on this potential value. The former aspect deals with dollar value per unit in the absence of pest damage and will vary from year to year, variety to variety and from orchard to orchard. This potential crop value may be affected by certain pest densities.

Actual crop value, that is the amount returned to the grower before his operating expenses, is determined by many factors outside the scope of this report but from a production standpoint the value is strongly influenced by (1) yield per land unit (usually per acre), (2) fruit quality, and (3) fruit size. In the following sections we will discuss each of these variables and the influence of pest damage on these value determinants.

Yield. Table 18 presents the variation in yield of 'Bartlett' pears over a 9-year period. This variation is more or less typical of all pear varieties grown in southern Oregon and was not pest induced, but rather the result of poor pollination or spring frosts.

As can be seen, total yield may vary enormously from year to year. When pest control decisions are made relative to this variable the potential yield will have to be estimated based on past history of particular orchards. This estimate will require constant updating as seasonal events raise or lower the yield estimates.

Other factors of fruit quality and fruit size being constant, an increase in the yield of fruit per acre will generally lower the economic injury level of a particular pest known to influence yield. This is due to the fact that control measures such as pesticides are applied on a per acre basis with associated costs related to acreage treated not to yield/acre. Thus, a miticide may cost $20/acre/application and based on a yield of 10 tons/acre this amounts to $2/ton. However, if yield were estimated at 20 tons/acre the cost is reduced to $1/ton. If mite density to be treated would have caused a $1.50/ton/acre reduction in crop yield the treatment would have been justified in the higher but not the lower producing situation.

Fruit quality. Table 19 summarizes the monetary return to growers received over a 6-year period for various grades of winter pears. In this table only returns for the U.S. Number 1 and Fancy grades are given as these represent the most common packs of fresh shipped pears from the Rogue River area. Lumping the returns for all varieties over a 6-year period (Table 19) the higher grade U.S. 1 averaged nearly 50% greater returns than the Fancy (U.S. 2) grade.

Over the years, improvements in fruit quality have been an important competitive market force in maximizing grower profits and has resulted in preferential prices being given for fruit which is nearly blemish free. Many of the quality standards now used will cause downgrading or culling of pears with no more than minor surface marking. These requirements have reduced the economic injury levels of pest species known to cause surface damage. The pear pests in southern Oregon important in determining fruit quality include the codling moth, pear psylla, San Jose scale, and the pear rust mite. Table 20 gives the acceptable level of damage for these pests for the various fruit grades.

The potential value of various fruit grades can be used in conjunction with potential pest control costs to set approximate levels of tolerable damage. Because of the high differential in monetary returns for the higher grades, the losses due to fruit marking often outweigh the cost of added control measures.

Table 18.
Variation in yield /acre of the Bartlett variety. 1964-1970. [a]

Year	Tons/Acre
1964	11.8
1965	10.8
1966	8.0
1967	12.3
1968	12.3
1969	14.2
1970	5.6

[a] Yield from Southern Oregon Experiment Station mature Bartlett pear orchard.

Table 19.
Effect of fruit grade on average $ return to grower for winter pear varieties. 1971-1976. [a]

Return	Variety and $ return per packed box for 2 grades					
	Bosc		Comice		D'Anjou	
1971-76	U.S. 1	Fancy	U.S. 1	Fancy	U.S. 1	Fancy
Average	4.30	2.68	4.80	1.61	2.70	2.02
Low	3.28	1.28	2.68	0.98	1.96	1.06
High	6.02	4.35	8.34	2.15	3.50	3.15

[a] Southern Oregon Experiment Station

For instance, using the average returns from Table 21, if pear psylla density on the 'Comice' variety was to reach levels that caused a 5% decrease in grade from U.S. 1 to U.S. 2 and if the yield was 10 tons/acre, the net loss would be about $70/acre. Under current conditions this situation would justify at least 2 additional pesticide treatments. However, on the 'D'Anjou' variety a similar decrease of 5% would lower crop value by only $15/acre currently not sufficient to offset additional pesticide treatment.

Fruit size. There is a definite tendency for the public to prefer pears of certain sizes. These will normally return premium prices. Table 22 presents the effect of fruit size on the value of four pear varieties over a 6-year period. As is

Table 20.
Acceptable damage for various winter pear fruit grades due to major pest species.

Pest	Fruit grade	Damage permitted
Codling moth	U.S. Extra fancy	2 healed slight stings.
	U.S. No. 1	2 healed stings.
	Fancy	3 healed stings.
Pear psylla	U.S. Extra fancy	Less than 10% surface with thinly scattered spotting.
		Less than 1/2 inch diameter with moderately scattered spotting.
		Less than 3/8 inch diameter with heavily concentrated spotting.
	U.S. No. 1	Less than 25% surface with thinly scattered spotting.
		Less than 3/4 inch diameter with moderately scattered spotting.
		Less than 1/2 inch diameter with heavily concentrated spotting.
	Fancy	Less than 1/2 inch of surface with thinly scattered spotting.
		Less than 1-1/4 inch diameter with moderately scattered spotting.
		Less than 3/4 inch diameter with heavily concentrated spotting.
San Jose Scale	U.S. Extra fancy	1 scale or scale mark.
	U.S. No. 1	2 scale or scale marks.
	Fancy	6 scale or scale marks.
	Export all grades	No more than 2% of lot with scale.
Pear rust mite on smooth skinned varieties	All U.S. grades	Smooth russett permitted on portion of calyx end not visible for more than 1/2 inch along the contour of pear when placed calyx end down on a flat surface.

evident there are substantial differences in dollar return for various sizes. In the case of 'Bosc' and 'D'Anjou' varieties there is a trend for lower value for the very largest and smallest sizes. With 'Bartlett' and 'Comice' a direct relationship exists between larger fruit and greater value.

The shifts in fruit size caused by insect or mite feeding are the result of indirect damage especially to leaf tissue that reduces the total amount of photosynthate. The feeding injury of several pests including the pear psylla, San Jose scale and the twospotted spider mite, appear to effect fruit size. However, except for the last pest species, studies have not been made on pears relating pest density to changes in fruit size.

Severe leaf damage may cause a shift to smaller fruit sizes and at times this will have important economic effects most years. In the case of the 'Bosc' variety there is a difference in return of $1.25/packed box between the 135 and the 150 fruit/box sizes (Table 22). Based again on 10 tons of fruit/acre yield (ca. 450 packed boxes) an increase in 10 boxes/acre from the 135 to the 150 size would reduce returns by about $12.50/acre or about the cost of one acaricide treatment.

Total crop value. It is apparent from the above discussion that three aspects of crop value (yield, quality and size) are highly variable even in the absence of pest damage. Since a known value is necessary to set economic injury levels for the various pest species, the question arises as to just what dollar figure is to be placed on a particular pear crop in a particular orchard in order to make appropriate decisions regarding control tactics. As an example, a study of Table 19 will reveal that for any of the three winter varieties evaluated, the value varied as much as $5/packed box between years with the high and low returns. To complicate the matter even more, the final value is not known until months after harvest and comes belatedly after all pest control decisions have been made for that particular year.

Table 21.
Average pear crop value (1971-1977)
of winter varieties from all sizes and grades. [a]

Variety and $ value per acre		
Bosc	Comice	D'Anjou
$1,700.	$1,850.	$1,100.

a Return to Southern Oregon Experiment Station, based on estimated average yield of 10 tons/acre with U.S. No. 1, Fancy, and cull fruit making up 85, 10, and 5% of the packout, respectively.

Table 22.
Effect of fruit size on $ return to grower
for winter pear varieties.
1971-76. U.S. No. 1 grade.

<table>
<tr><th></th><th colspan="11">Fruit Sizes</th></tr>
<tr><th>Variety</th><th>60</th><th>70</th><th>80</th><th>90</th><th>100</th><th>110</th><th>120</th><th>135</th><th>150</th><th>165</th><th>180</th></tr>
<tr><td>Bosc</td><td>$3.30 [a]</td><td>4.24</td><td>4.98</td><td>5.42</td><td>5.34</td><td>5.42</td><td>5.35</td><td>5.24</td><td>3.99</td><td>3.01</td><td>2.03</td></tr>
<tr><td>D'Anjou</td><td>2.59</td><td>2.78</td><td>3.07</td><td>3.08</td><td>3.06</td><td>3.05</td><td>3.00</td><td>2.88</td><td>2.42</td><td>2.21</td><td>1.41</td></tr>
<tr><td>Comice</td><td colspan="4">6.02</td><td colspan="2">5.64</td><td colspan="2">4.24</td><td colspan="2">0.93</td><td></td></tr>
</table>

[a] Price per packed box.

Table 23.
1977 spray program for pear pests in southern Oregon.

Timing of application	Material and approximate cost ($) per acre		Pests at which treatment directed
Dormant	Oil	8.00	Pear psylla
Late Dormant	Perthane	24.00	Pear psylla
	+		
	Diazinon	16.00	San Jose scale
	+		
	Oil	5.00	Pear psylla, San Jose scale, mite eggs
Pink	Morestan	20.00	Pear psylla, rust mite
1st Summer	Thiodan	18.00	Pear psylla
	Guthion	6.00	Codling moth
2nd Summer	BAAM	30.00	Pear psylla
	Guthion	6.00	Codling moth
	Plictran	20.00	Spider mites
	Diazinon	16.00	San Jose scale
3rd Summer	BAAM	30.00	Pear psylla
	Guthion	6.00	Codling moth
	Plictran	20.00	Spider mites
	Total	$217.00	

Cost for individual pests	
Pear psylla	$120.
Pear rust mite	10.
Spider mites	42.
San Jose scale	32.
Codling moth	18.

Though growers are apt to disagree, it is probably advised to use average crop value rather than high or low average values in assessing the damage potential of pest species. Using for a standard value the maximum crop value received in the past will lower economic injury thresholds to a point when unnecessary control measures become routine. On the other hand, using as an economic base the low return will most often result in a higher incidence of unacceptable pest damage.

Average crop value can usually be calculated based on packing house printouts for varieties within these orchards. As mentioned previously the estimate of crop value will have to be updated with seasonal events that influence yield, quality or fruit size.

Crop value and current costs for pest control. The current costs for pest control (Table 23) can be compared to potential crop value to arrive at a very general level of permissible damage by particular pests (Table 24). If, for example pear psylla control costing $120/acre is compared to the average value of the three winter varieties, we find it amounts to 7%, 6% and 11% of the average value for 'Bosc', 'Comice' and 'D'Anjou', respectively. Thus, depending upon the pear variety, if no chemical control program was used for this pest the grower could afford from 6-11% crop loss and not have incurred economic loss. An increase in crop value or a decrease in control costs will lower the tolerable damage level. The work of IPM is to relate potential damage by various pest densities to the fluctuating crop values described above and to develop control tactics which are realistic for the grower in the economic paradigm.

Horticultural Aspects

As horticultural practices influence either crop value, pest density or pest damage expression, they will influence the practice of integrated pest management. The horticultural aspects which are most pertinent to IPM include (1) the

Table 24.
Current cost of control for various pear pests as expressed by % of crop value for winter pear varieties.

	Variety		
Pest species	Bosc	Comice	D'Anjou
Codling moth	1.1%	1.0%	1.6%
Pear psylla	7.1	6.4	10.9
San Jose scale	1.9	1.7	2.9
Pear rust mite	0.6	0.5	0.9
Spider mites	2.5	2.3	3.8

selection of appropriate varieties and understocks and (2) the development of horticultural practices that maximize or stablize production of high quality pears or those that are implemented to reduce production costs.

Pear varieties. Of the 300-400 pear varieties described only four or five are of commercial significance in southern Oregon. Each has its own characteristics including variations in susceptibility to pest attack. There are also variations between varieties in their value. Crop value determines the economic impact of pest damage and has been covered in the first part of this section.

Variation of the pear varieties to injury has been shown with several pear pests including the codling moth and the pear psylla, and is probably present with most other pear pests. There are several reasons why certain varieties of pear are able to tolerate higher pest densities than others. One example deals with the difference in damage caused by pear psylla honeydew which is more noticeable on the clear skinned varieties, such as 'D'Anjou' or 'Comice', than on the normally dark russetted 'Bosc' variety. This allows for a somewhat higher tolerance to psylla densities on the latter variety. Also, a pear variety may be more tolerant to pest attack due to various morphological or physiological attributes such as thicker leaf cuticle (spider mites) or the presence of stone cells in the fruit (codling moth). Finally, variation in tree vigor differs among varieties and may influence population density of so-called "flushfeeders" such as pear psylla and the pear rust mite.

All pear varieties are grafted onto understocks of other *Pyrus* (occassionally quince) varieties or species. These stocks are selected for several purposes including resistance to root insects or disease, to promote precocious bearing, for tree growth control or to maximize yields. The influence of rootstocks on yield can be dramatic. Work in southern Oregon (Lombard and Westwood 1976) showed that there may be as much as a 3-4 fold increase in pear tree productivity depending upon the rootstock used. Orchards planted using the more efficient rootstocks will have increased yields and this will tend to lower the economic injury levels for many pest species. The rootstock may also influence pest levels on the top worked varieties. Though few studies have been conducted to quantify these effects a significant difference was measured between population densities of the pear rust mite on 'Bartlett' pears top worked to various rootstocks (Table 25).

Horticultural practices. In addition to the choice of rootstock and variety there are a number of horticultural practices which either enhance or decrease the potential for IPM implementation. A few of these practices are given in Table 26 along with their apparent effects on IPM.

It is also apparent that some of the newer developments in horticulture will alter the current crop economics which in turn will influence the management of

Table 25.
Influence of understocks on population density of pear rust mite on the Bartlett cultivar.

Understocks	Rust mites/leaf
Old Home clonal	81.8 a
Old Home/Quince A	82.5 a
Nivalis	82.5 a
Old Home/Bartlett seedling	83.4 ab
Old Home/Nelis	90.1 abc
Calleryana	100.5 bc
Nelis seedling	105.6 c
Bartlett seedling	106.2 c

Table 26.
A partial list of various horticultural practices and their effects on integrated pest management of pear pests.

Practice	Effect on IPM
Irrigation (furrow or under tree sprinkler	Due to slow drying conditions on heavy soils, prevents currative spray practices and leads to preventative pesticide applications.
Irrigation (overtree)	a) Removes pesticide deposits and may cause increases in frequency of chemical treatments (see codling moth). b) Reduces injury and/or population density of pest species (see pear psylla, pear rust mite and spider mites).
Pruning (mechanical tree topping)	Promotes new growth which may cause increases in pest levels (see pear psylla).
Post harvest defoliation	To allow early pruning. May decrease overwintering populations of some pest levels (see spider mites).
Fertilizer treatment	No studies available from southern Oregon, but excess nitrogen applications have been reported to result in increases in densities of several orchard pest species

pear pests. One of the recent trends in orchard management is to replace the older orchards with high density plantings of over 500 trees/acre. Yields from these plantings are expected to be 2-3 times that from the older standard planted orchards and will have the effect of reducing economic injury levels and treatment thresholds accordingly.

REFERENCES

Barnes, M. M. 1959. Deciduous fruit insects and their control. *Ann. Rev. Entomol* **4**:343-362.

Batiste, W. C., W. H. Olsen, and A. Berlowitz. 1973. Codling moth: influence of temperature and daylight intensity on periodicity of daily flight in the field. *J. Econ. Entomol.* **66**:883-892.

Batjer, L. P., and H. Schneider. 1960. Relation of pear decline to rootstocks and sieve-tube necrosis. *Proc. Amer. Soc. Hort. Sci.* **76**:85-97.

Brunner, J. F. 1975. Economic injury level of the pear psylla, *Psylla pyricola* Foerster, and a discrete time model of a pear psylla-predator interaction. Ph.D. Dissertation. Washington State University, Pullman. 94 pp.

Burts, E. C. 1968. An area control program for the pear psylla. *J. Econ. Entomol.* **61**:261-263.

Burts, E. C., and A. H. Retan. 1973. Detection of pear psylla. *Wash. State Univ. E.M.* 3069. 2 pp.

Cordy, C. B. 1977. History of the Rogue Valley fruit industry. mimeo. 30 pp.

Downing, R. S., and D. M. Logan. 1977. A new approach to San Jose scale control. (Hemiptera: Diaspididae). *Can. Entomol.* **109**:1249-1252.

Gentile, A. G., and F. M. Summers. 1958. The biology of San Jose scale on peaches with special reference to the behavior of males and females. *Hilgardia* **27**:269-285.

Hoyt, S. C., and E. C. Burts. 1974. Integrated control of fruit pests. *Ann. Rev. Entomol.* **19**:231-252.

Lombard, P. B., P. H. Westigard, and D. Carpenter. 1966. Overhead sprinkler system for environmental control and pesticide application in pear orchards. *HortScience* **1**:95-96.

Lombard, P. B., and M. N. Westwood. 1976. Performance of six pear cultivars on clonal Old Home, double-rooted and seedling rootstocks. *Jour. Amer. Soc. Hort. Sci.* **101**:214-216.

Madsen, H. F., and C. V. G. Morgan. 1970. Pome fruit pests and their control. *Ann. Rev. Entomol.* **15**:295-320.

Madsen, H. F., and J. M. Vakenti. 1973. Codling moth: use of codlemone baited traps and visual detection of entries to determine need of sprays. *Environ. Entomol.* **2**:677-679.

McMullen, R. D., and C. Jong. 1967. New records and discussion of predators of the pear psylla, *Psylla pyricola* Foerster, in British Columbia. *J. Entomol. Soc. British Columbia.* **64**:35-40.

Pickel, C. 1976. Computerized forecasting as an aid for timing codling moth control. MS Thesis, University of California, Berkeley. 37 pp.

Riedl, H., B. A. Croft, and A. J. Howitt. 1976. Forecasting codling moth phenology based on pheromone trap catches and physiological-time models. *Can. Entomol.* **108**:449-460.

Stern, V. M., R. F. Smith, R. van den Bosch, and K. S. Hagen. 1959. The integration of chemical and biological control of the spotted alfalfa aphid. Part I. The integrated control concept. *Hilgardia* **29(2)**:81-101.

Terriere, L. C., U. Kiigemagi, R. W. Zwick, and P. H. Westigard. 1966. Persistance of pesticides in orchards and orchard soils. *Adv. Chem. Ser.* **60**:263-270.

Westigard, P. H. 1969. Timing and evaluation of pesticides for control of the pear rust mite. *J. Econ. Entomol.* **62**:1158-1160.

Westigard, P. H. 1970. Unpublished data. Medford, Oregon.

Westigard, P. H. 1971. Integrated control of spider mites on pear. *J. Econ. Entomol.* **64**:496-501.

Westigard, P. H. 1973a. The biology of and effect of pesticides on *Deraeocoris brevis piceatus* (Heteroptera: Miridae). *Can. Entomol.* **105**:1105-1111.

Westigard, P. H. 1973b. Pest status of insects and mites on pear in southern Oregon. *J. Econ. Entomol.* **66**:227-232.

Westigard, P. H. 1975. Population injury levels and sampling of the pear rust mite on pears in southern Oregon. *J. Econ. Entomol.* **68**:786-790.

Westigard, P. H. 1977. San Jose scale control on pears in southern Oregon. *Ore. Hort. Soc.* **68**:44-47.

Westigard, P. H. 1978. Unpublished data. Medford, Oregon.

Westigard, P. H., and D. W. Berry. 1964. Control of the pear rust mite, *Epitrimerus pyri. J. Econ. Entomol.* **57**:953-955.

Westigard, P. H., and D. W. Berry. 1970. Life history and control of the yellow spider mite on pear in southern Oregon. *J. Econ. Entomol.* **63**:1433-1437.

Westigard, P. H., and L. D. Calvin. 1971. Estimating mite populations in southern Oregon pear orchards. *Can. Entomol.* **103**:67-71.

Westigard, P. H., and L. D. Calvin. 1977. Sampling San Jose scale in a pest management program on pear in southern Oregon. *J. Econ. Entomol.* **70**:138-140.

Westigard, P. H., and K. L. Graves. 1976. Evaluation of pheromone baited traps in a pest management program on pears for codling moth control. *Can. Entomol.* **108**:379-382.

Westigard, P. H., and R. W. Zwick. 1972. The pear psylla in Oregon. *Ore. Agr. Exp. Sta. Tech. Bull.* 122. 22 pp.

Westigard, P. H., L. G. Gentner, and D. W. Berry. 1968. Present status of biological control of the pear psylla in southern Oregon. *J. Econ. Entomol.* **61**:740-743.

Westigard, P. H., L. Gentner, and B. A. Butt. 1975. Codling Moth: Egg and first instar mortality on pear with special reference to varietal susceptibility. *Environ. Entomol.* **5**:51-54.

Westigard, P. H., U. Kiigemagi, and P. B. Lombard. 1974. Reduction of pesticide deposits on pear following overtree irrigation. *HortScience* **9**:34-35.

Westigard, P. H., P. B. Lombard, and D. W. Berry. 1967. Bionomics and control of the twospotted spider mite on pear in southern Oregon. *Ore. Agr. Exp. Sta. Tech. Bull.* 101. 32 pp.

Westigard, P. H., P. B. Lombard, and J. H. Grim. 1966. Preliminary investigations of the effect of feeding of various levels of the twospotted mite on its D'Anjou pear host. *Proc. Amer. Soc. Hort. Sci.* **89**:117-122.

Westigard, P. H., L. E. Medinger, and O. E. Kellogg. 1972. Field evaluation of pesticides for their suitability in an integrated program for spider mites on pear. *J. Econ. Entomol.* **65**:191-192.

Westigard, P. H., M. N. Westwood, and P. B. Lombard. 1970. Host preference and resistance of *Pyrus* species to the pear psylla, *Psylla pyricola* Foerster. *J. Amer. Soc. Hort. Sci.* **95**:34-36.

Westwood, M. N., H. R. Cameron, P. B. Lombard, and C. B. Cordy. 1971. Effects of trunk and rootstock on decline, growth and performance of pear. *J. Amer. Soc. Hort. Sci.* **96**:147-150.

Yothers, M. A., and E. R. VanLeeuwen. 1931. Life history of the codling moth in the Rogue River Valley of Oregon. *USDA Tech. Bull.* 255. 35 pp.

Zwick, R. W., and P. H. Westigard. 1978. Prebloom petroleum oil applications for delaying pear psylla (Homoptera: Psyllidae) oviposition. *Can. Entomol.* **110**:225-236.

INTEGRATED PEST MANAGEMENT SYSTEMS IN PENNSYLVANIA APPLE ORCHARDS

Dean Asquith and Larry A. Hull
Pennsylvania State University
Fruit Research Laboratory
Biglerville, Pennsylvania

INTRODUCTION

Integrated pest management systems that achieve economic control of the European red mite, *Panonychus ulmi* (Koch), and the twospotted spider mite, *Tetranychus urticae* Koch, by the predator *Stethorus punctum* (LeConte) while holding injury by insect pests below economic levels with applications of reduced amounts of insecticides are viable, successful management techniques practiced by Pennsylvania apple growers. Development of these management systems required much painstaking research as well as trial and error in actual orchard situations.

During the 1960's control of the European red mite with pesticides became extremely frustrating and expensive. No matter how efficient an acaricide was in controlling the European red mite, when the chemical was introduced into orchard practice, strains of the European red mite became resistant to it within a period of two to five years. Selecting effective acaricides suitable for orchard use became an unending battle. In view of this situation, success in controlling populations of mites by biological means while at the same time holding populations of insect pests below economic injury levels as accomplished in Nova Scotia by Pickett (1959) and coworkers (Pickett *et al.* 1956) and in Washington state by Hoyt (1969) and coworkers encouraged the senior author to investigate the possibilities of controlling mites by the integrated method in Pennsylvania apple orchards. Two steps were taken in 1966 that led to prompt but limited progress in these investigations. First, an orchard was set aside and operated as an environment for predators. Second, Robert L. Horsburgh was hired as a research assistant and graduate student because he had extensive experience as an extension entomologist with responsibility for the operation of a Nova Scotia integrated pest management system. Within the next year a grant was obtained from the ARS which helped to support the integrated pest management project financially.

INITIAL INVESTIGATIONS

During 1967, his first year in the field, Horsburgh had two prime responsibilities: (1) to conduct a survey of predators of the European red mite and the twospotted spider mite in commercial and abandoned apple orchards; and (2) to observe closely the seasonal development of insect and mite populations in our so-called biological control apple orchard. He found more than 30 predators of mites in Pennsylvania apple orchards (Horsburgh and Asquith 1968). In the meantime in our biological control orchard, which was being treated with reduced amounts of pesticides, Horsburgh discovered *Hyaliodes vitripennis* (Say), a very efficient predator of the European red and twospotted spider mites. However, he soon learned that *H. vitripennis* was highly susceptible to the insecticides commonly employed in orchard practice. Also, when he attempted to select a strain with resistance to a particular chemical such as azinphosmethyl, he found that any degree of resistance was associated with some loss of reproductive powers.

Although Horsburgh continued to study *H. vitripennis*, a decision was made to place several apple orchards on a schedule of reduced amounts of pesticides to determine if one or more predators might assert themselves in this type of situation. Several growers agreed to cooperate and their apple orchards were placed on spray schedules involving application of reduced amounts of pesticides, especially insecticides. In five of the six orchards selected, European red mite populations began to build in mid-summer and very shortly after this increase started, populations of *S. punctum* moved into the trees, fed on European red mites and on twospotted spider mites if they were present. The predators soon began to reproduce, with female *S. punctum* depositing eggs on apple leaves. Both *S. punctum* adults and larvae feed on mites (Figs. 1 and 2). The adults can fly and therefore move about the orchard locating the heaviest mite populations before they can cause economic injury to the trees. On the other hand, larvae cannot fly and are confined to the tree on which they hatch from eggs. The larvae, therefore, are forced to clean up populations of mites on individual trees in order to survive whereas the adults can move to other trees when populations of mites are reduced to extremely low levels.

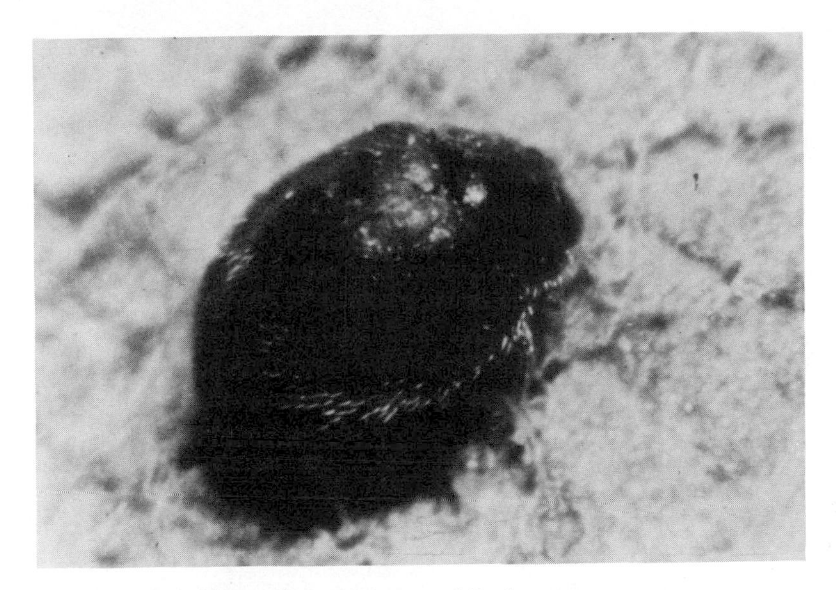

Figure 1. Adult stage of *Stethorus punctum.*

Figure 2. Larval stage of *Stethorus punctum.*

In southern Pennsylvania, *S. punctum* produces three generations a year. The species overwinters as adults, some of them beneath the duff underneath apple trees, others in protected areas such as fence rows and brushy sections of nearby woodlands. In the spring, adults emerge from winter quarters during warm, sunny days and begin feeding on mites and mite eggs on the trunks. They move into the trees when apple buds are in the pink stage of development and continue into the early post-bloom period. The movement into the trees coincides with hatching of European red mite eggs. Since these adults have not fed since the previous fall, they are somewhat weakened and are not as resistant to insecticides as are members of the summertime generations. Richard B. Colburn joined the entomology group in the fall of 1968 and through his endeavors contributed much information on the morphology and biology of *S. punctum* (Colburn and Asquith 1971).

SELECTIVE PESTICIDE PROGRAM

A phase of the integrated pest management project of prime importance was the selection of insecticides and their respective dosages that *S. punctum* populations could tolerate and at the same time could be employed to regulate populations of damaging insect pests below economic injury levels. In Table 1 will be found lists of fungicides, insecticides, and acaricides that through field testing were found compatible with the "Pennsylvania Integrated Pest Management System for Apples" when used at specified per acre dosages. Details of the tests in which many of the listed pesticides were selected for suitability in the integrated program may be found in the following publications: Asquith and Hull 1973; Asquith *et al.* 1976a, b, c, d, e, f; Asquith *et al.* 1977a, b, c.

Selection of pesticides for use in an integrated pest management system has to be continuous. In the developmental stage new pesticides have to be screened for at least three reasons important to the apple integrated pest management approach. Initially, they should be screened for effectiveness on the many pests that attack the apple and also to determine whether they are phytotoxic to any part of an apple tree. Concurrently, new pesticides should be tested for tolerance by *S. punctum* as both larvae and adults. Such testing is not easy and requires careful observation, collection of data and analysis of data by research entomologists. Such analyses should permit entomologists to evaluate new pesticides and decide whether they merit extensive orchard trials.

In Pennsylvania it has been found necessary to submit new pesticides that have passed the initial replicated, comparatively small scale tests, to orchard size trials. In these trials new pesticides are mixed in the spray tank with other pesticides that have to be used at the same time and are applied as 6X

Table 1.
Pesticides compatible with the Pennsylvania integrated pest management program for apples.

Fungicides[a]	Insecticides	Acaricides
Benlate (benomyl)	Guthion (azinphosmethyl)	Carzol (formetanate)
Captan	Imidan (phosmet)	Plictran (cyhexatin)
Cyprex (dodine)	Zolone (phosalone)	Vendex
Polyram	Penncap M (methyl parathion)	Omite (propargite)
Dikar (dinocap)	Cygon (dimethoate)	
Karathane (dinocap)	Systox (demeton)	
Sulfur	Lannate (methomyl)	
	Thiodan (endosulfan)	
	Superior Oil	

[a]EC formulations are less satisfactory than wettable powders. It is necessary with many pesticides to use specifically selected dosages. [Modified from: Asquith (1979).]

concentrates by the alternate middle technique of spraying (Lewis and Hickey 1967; Asquith and Colburn 1971). Such orchard size trials make it possible to determine whether the per acre dosage of the new pesticide chosen is sufficient to hold the pest species below economic injury levels and if that dosage can be tolerated by *S. punctum* larvae and adults in practical orchard situations. During the course of such trials, the investigator sometimes learns that mixing the new pesticide with certain other pesticides may cause injury to parts of the apple tree. In cases like this, satisfactory adjustments can sometimes be made and in other cases the matter is so serious that the new pesticide has to be dropped. Orchard evaluations of this kind have to be made carefully and usually have to be repeated to be sure of their accuracy. Careful orchard investigations with pesticides has enabled research entomologists in Pennsylvania to introduce several new pesticides into the integrated pest management system. They have also been able to manipulate the system to utilize both old and new pesticides to control new pests without greatly disrupting the integrated system. The pest requiring the greatest amount of manipulation has been the tufted apple budmoth, *Platynota idaeusalis* (Walker). Successful selection of pesticides to control a resurgence of populations of the white apple leafhopper, *Typhlocyba pomaria* McAtee, without disruption of the integrated pest management system is the kind of results research entomologists have to attain (Asquith *et al.* 1975). In the last two seasons, the rosy apple aphid, *Dysaphis plantaginea* (Passerini), has required careful selection of pesticides and manipulation of the pest management system.

During the first stages of implementation of the integrated pest management system in Pennsylvania apple orchards, entomologists did not know to what degree populations of *S. punctum* could be relied upon to hold populations of the European red mite below economic injury levels. For this reason, the early season (first half) of the spray schedule portion of the integrated program relied upon the pesticide dinocap in either the product Karathane® or Dikar® to suppress infestations of the European red mite until populations of *S. punctum* attained sufficient numbers to contain the European red mite infestations. Since dinocap is still the most commonly employed fungicide for control of powdery mildew *(Podosphaera leucotricha)* in Pennsylvania apple orchards, many apple growers continue to follow this practice. However, field trials in Pennsylvania have shown that in many cases where dinocap is not needed for control of the powdery mildew fungus, it is also not required to suppress growth in the population of European red mites. About one-half the time when dinocap is not present in the spray schedule, no acaricide is required to suppress the European red mite population. In other situations, application of a low dosage of acaricide may be required to suppress the mite population in order to permit the population of *S. punctum* to attain satisfactory numbers (Hull *et al.* 1978) (Fig. 3). It has been the experience of the authors that once a population of *S. punctum* has attained sufficient numbers to regulate populations of the

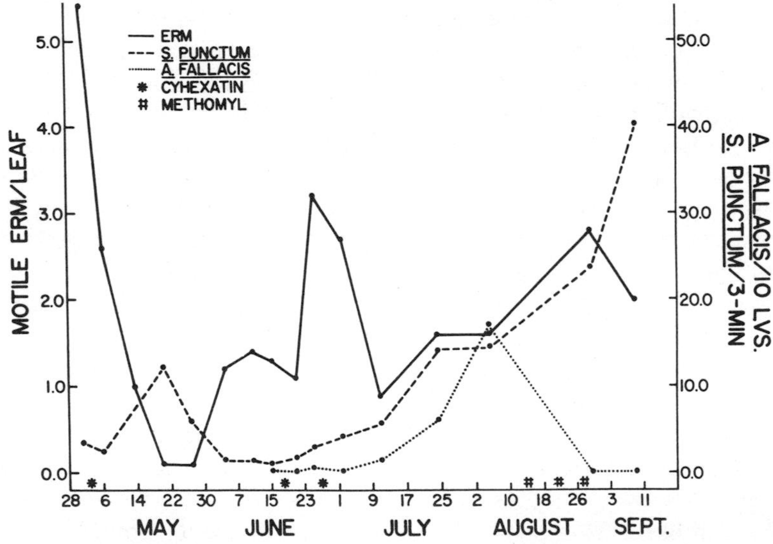

Figure 3. Seasonal history of the European red mite, *S. punctum* and *A. fallacis* in the Arendtsville Experiment, where no oil or dinocap was applied. [Taken from: Hull *et al.* (1978)].

European red mite below economic injury levels, the populations of European red mites do not again, during that season, attain high enough numbers to cause economic damage.

In establishing and maintaining the "Pennsylvania Integrated Pest Management System for Apples," selection of pesticides to regulate populations of pests below economic injury levels and for tolerance by *S. punctum* has been a primary research effort. Also, determining the influence of pesticide applications on the seasonal development of populations of *S. punctum* and the need to suppress or not suppress mite infestations with chemicals has contributed to the success of the integrated program. To adapt this program to the systems analysis approach to pest management, some sophisticated biological and statistical investigations have been necessary.

SYSTEMS ANALYSIS APPROACH

In preparing to utilize the systems analysis approach to aid decision-making in the Pennsylvania integrated pest management system, acquisition of large quantities of data and validation of monitoring methods were essential. The effort to construct a simulation model of the interaction between the European red mite and *S. punctum* and the orchard environment helped direct the search for relevant data.

Among the information necessary for the construction of a simulation model of the interaction between the European red mite and *S. punctum* was the development of the European red mite in relation to temperature. The results of the research of Cagle (1946) proved adequate for this purpose. For years in Pennsylvania, leaf counts of the European red mite on samples of 20 leaves taken at breast to head height around sample trees had proved adequate for establishing the rise or decline of mite populations on trees in tests of acaricides. Regular counts of mites by this method proved satisfactory on-the-spot data for the integrated pest management system.

A reliable method of monitoring populations of *S. punctum* larvae and adults was required to supply data for the simulation model. During the development of the integrated pest management system, *S. punctum* populations were estimated by trained observers, counter in hand, walking slowly around the periphery of an apple tree and counting adult and larval *S. punctum* that fell within their range of vision in a period of 3 minutes (Asquith and Colburn 1971). In conjunction with sampling of *S. punctum,* two experiments were conducted in Pennsylvania.

Experiment I

This experiment was designed to relate the number of *S. punctum* predators counted in a 3-min. observation and the number per 1,000 leaves on an apple tree. This information was needed because the 3-min. observation, widely used in pest management applications, does not estimate the absolute predator population needed for computer modeling purposes. The experiment consisted of sampling each of 10 trees using the 3-min. observation and then estimating the number of predators/1,000 leaves using a beating sample. Using regression analysis, the functional relationship between the number of predators/1,000 leaves and the 3-min. observation was estimated for adult and larval populations. The predictive equations and the R-squared values are given below. For modeling purposes, these predictions are combined with an estimate of the number of leaves/tree to estimate the absolute predator population/tree.

ADULTS $Y = 14.3 + 0.0047x^2$ R-SQUARED = 63%

LARVAE $Y = 0.862x$ R-SQUARED = 89%

Where Y equals the number of *S. punctum*/1,000 leaves, and x equals the number of *S. punctum*/3-min. observation.

Experiment II

This experiment was designed to investigate the accuracy of the 3-min. observation, the commonly used method of counting *S. punctum* in Pennsylvania apple orchards. Two hypotheses were tested: (1) There is no real difference between human observers and the number of *S. punctum*/3-min. observation when the same tree is counted by different observers; and (2) the number of *S. punctum*/3-min. observation is directly proportional to the number per tree. The first hypothesis investigates the possibility that the 3-min. observation may be biased by a tendency for different observers to systematically count more or less predators. The second is important because it is very desirable to use a population index in which, *e.g.*, twice as many predators/-3-min. observation means twice as many predators/tree. The experiment was conducted by counting on each of 4 trees consecutively by 3 different trained observers using the 3-min. observation. Data for each tree was then taken using a more accurate 8-min. section count. The results were analyzed using an analysis of covariance model in which the covariate was the 8-min. count and the AOV factor was the observer. No statistically significant difference was found between observers. A straight line through the origin accurately described the relationship between the number of *S. punctum* adults and larvae/3-min. observation and the number/8-min. count, suggesting the validity of the second hypothesis.

OTHER BIOLOGICAL MONITORING

While regular monitoring of the European red mite and *S. punctum* populations is essential to the construction of a simulation model of the interaction between the European red mite and *S. punctum* and the orchard environment, data on other factors are of prime importance. Records of daily minimum and maximum temperatures play an important part in tracing the development of European red mite populations. Also precise records on the use of pesticides, especially acaricides, are important in assessing sudden shifts in both European red mite and *S. punctum* populations. Since chemical pesticides are required for regulation of populations of pests other than mites, in order to improve the efficiency of pesticide applications, it is important to monitor regularly the status of all pest populations in the orchard environment. In Pennsylvania, sex pheromone traps of the codling moth, *Laspeyresia pomonella* (L.), the redbanded leafroller, *Argyrotaenia velutinana* (Walker), the oriental fruitmoth, *Grapholitha molesta* (Busck), the tufted apple budmoth, *Platynota idaeusalis* (Walker), and the variegated leafroller, *Platynota flavedana* (Clemens) have been employed to monitor population shifts of these pests. Synthetic pheromones attractive to the latter two insects were developed by chemists at the New York State Agricultural Experiment Station (Hill *et al.* 1974 and 1977). In Pennsylvania, W. M. Bode field-evaluated the sex pheromone mixtures for effectiveness in attracting males of *P. idaeusalis* and *P. flavedana.* Besides evaluating formulations of the sex attractants, Bode compared the effectiveness of two types of commercial adhesive-coated insect traps baited with the sex attractants for capturing male tufted apple budmoths and redbanded leafroller moths (Bode *et al.* 1973).

Three studies by Hull *et al.* (1976 and 1977a, b) contributed greatly to an understanding of the important function of *S. punctum* as the main predator of the European red mite in Pennsylvania. In the first piece of research on the mite searching ability of *S. punctum* within an apple orchard, the adults demonstrated the ability to find the European red mite even when their populations averaged less than one mite per leaf and most of the trees were kept free of mites (Hull *et al.* 1977b). *S. punctum* adults also demonstrated the ability to move throughout the orchard responding to where the mite populations were the highest and moving from trees where the European red mites had dwindled to low numbers. Once they found these trees, *S. punctum* further maximized their efficiency as predators by responding to areas of the tree where the mite populations were the highest (Fig. 4). The European red mite populations increased faster within the tree row where the leaves were not as heavily covered with spray material. This response of *S. punctum* populations to the changing

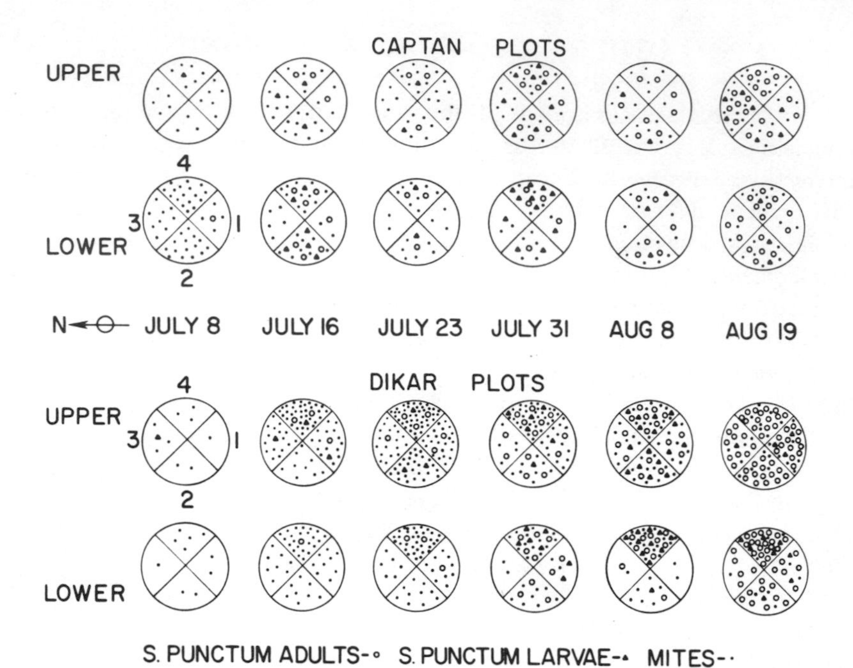

Figure 4. Distributional patterns of *Stethorus punctum* adults and larvae with respect to its prey, *Panonychus ulmi* within apple trees. Each circle represents the number of *S. punctum* adults and larvae (per 1-min. observation) and *P. ulmi* (per leaf) on each section of a Stayman apple tree. [Taken from: Hull *et al.* (1976)].

distribution of a European red mite population has been very important to the success of the Pennsylvania integrated pest management system because the female predators deposit eggs on apple leaves where prey are present.

In the third experiment, the functional response of the predator *S. punctum* was tested at various densities of the European red mite. Overwintered *S. punctum* adult females were tested at a range of densities from 4-30 mites/cage, whereas, the second and third generation larvae, female and male adults were tested at a range of densities from 4-80 mites/cage. In the spring test, the consumption rate of the overwintered *S. punctum* adult females increased as prey density increased from 4 to 30 (Table 2). The consumption rate of the second and third generations of *S. punctum* also increased with increasing prey density (Table 3). The adult female *S. punctum* consumed more mites per hour than the larvae and the larvae consumed more than the adult male *S. punctum.*

Table 2.
Mean number of European red mite adult females eaten/hour (±standard error) by overwintered *S. punctum* adult females at a mean temperature of 26.2°C (range 13-33).[a]

Prey density	Prey eaten/hour
4	1.38 ± 0.16
12	2.92 ± 0.21
16	4.42 ± 0.36
20	4.63 ± 0.47
30	5.32 ± 0.23

[a] Taken from: Hull *et al.* (1977).

Temperature also influenced the feeding rate of *S. punctum.* As temperatures increased during the day, it was paralleled by an increase in the feeding rate of *S. punctum.* Analyses of the data will be found in the paper of Hull *et al.* (1977a).

Growers, consultants and orchard managers using the integrated pest management method in Pennsylvania are constantly confronted with the problem of separating pest population complexes as they interact with various environmental factors. Through the recent development of systems analysis techniques these pest populations coupled with environment factors can now be studied as discrete components, thereby allowing the determination of how various pest management tactics will affect individual components. Recognizing the complex problems associated with integrated pest management systems, the research entomologists at The Pennsylvania State University Fruit Research Laboratory began a systems analysis modeling program in 1973.

COMPUTER SIMULATION OF PREDATOR-PREY INTERACTIONS

The first objective was to develop a prototype computer simulation model of the interaction between the European red mite and the predator *S. punctum.* This predator-prey interaction was chosen because (1) European red mites have been a serious pest of apple trees in Pennsylvania year after year; (2) *S. punctum* populations were providing good control of the European red mite throughout

Table 3

Mean number of European red mite adult females eaten/hour (±standard error) by 2nd and 3rd generation *S. punctum* female and male adults and larvae at various prey densities and temperatures.[a]

	Female adults			Male adults			Larvae		
Prey density	No. mites eaten	Mean temp. °C	Temp. range	No. mites eaten	Mean temp. °C	Temp. range	No. mites eaten	Mean temp. °C	Temp. range
4	1.39 ± 0.13	27.70	15-35	1.12 ± 0.47	27.74	18-32	1.13 ± 0.39	29.95	16-37
8	2.00 ± 0.57	27.70	15-35	1.68 ± 0.11	27.74	18-32	2.41 ± 0.61	29.95	16-37
12	3.56 ± 0.26	27.70	15-35	1.73 ± 0.52	27.74	18-32	2.25 ± 0.48	29.95	16-37
16	4.13 ± 0.24	27.70	15-35	2.45 ± 1.36	27.74	18-32	2.58 ± 0.44	29.95	16-37
20	2.69 ± 0.33	26.96	15-34	2.35 ± 0.16	28.31	19-33	3.63 ± 0.58	26.96	15-34
50	7.12 ± 1.11	26.96	15-34	4.42 ± 0.47	28.31	19-33	4.42 ± 0.61	26.96	15-34
80	10.44 ± 1.09	26.96	15-34	5.52 ± 0.26	28.31	19-33	7.10 ± 0.86	26.96	15-34

[a] Taken from: Hull *et al.* (1977).

Pennsylvania apple orchards; (3) *S. punctum* were tolerant to many of the pesticides currently recommended in Pennsylvania (Asquith and Hull 1973); and (4) the successful management of this predator-prey interaction required a consideration of many individual components.

The computer simulation model considers the European red mite population on a tree to be a system composed of components representing the various life stages (Fig. 5). The three intrinsic components of the model are summer egg incubation, the rate of development of newly hatched mite larvae up through the deutonymph stage, and adult mite development and oviposition. The extrinsic components of this system which affect the population dynamics of the mite are composed of orchard temperatures, the *S. punctum* population, and the orchard spray program. The rate of growth and development of the European red mite is primarily determined by temperature and the developmental rates used in the model are based on the work of Cagle (1946). The temperatures which are needed in the model are generated from a normal distribution with mean and variance based on Pennsylvania's midrange temperature records for that date (Mowery *et al.* 1975).

After the prototype computer simulation model was constructed, the decision was made to build a predictive system, later termed MITESIM (Mowery *et al.* 1977), that would assist growers, consultants and orchard managers in making appropriate decisions concerning the management of European red mite populations. The primary objective of MITESIM is to maximize the predation abilities of *S. punctum* to mitigate populations of the European red mite below economic injury levels. The system makes decisions on when predation by *S. punctum* needs to be supplemented by a small dosage of a miticide to assist the predator in maintaining the mite population at a low level. Due to the mobility of *S. punctum* and its inherent ability to locate increasing mite populations in

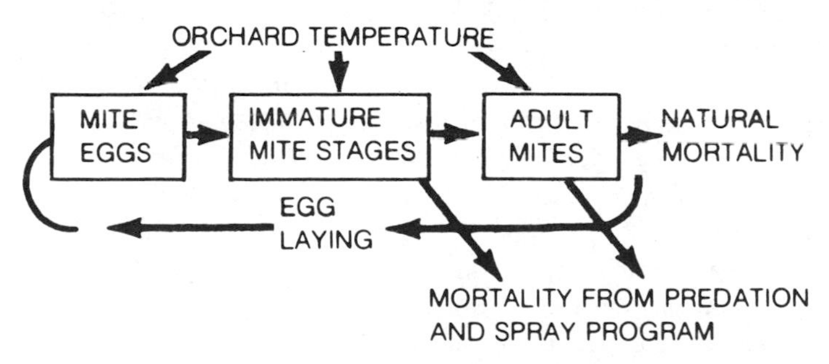

Figure 5. Schematic drawing showing how the European red mite population in an orchard, and the factors affecting it, are represented in the computer simulation model. [Taken from: Mowery and Asquith (1975)].

apple orchards, a miticide should be used only to prevent the European red mite from causing damage to the trees. Overuse of miticides by the grower will deplete the mites as a food supply for *S. punctum,* increase the probability of the mites developing resistance, increase the cost of control, and pollute the environment unnecessarily. MITESIM is a system designed to lessen the chance for the occurrence of these factors.

For a grower or pest management consultant to use MITESIM, four pieces of information must be gathered from the orchard to be monitored. First, an initial count of the mite life stages must be made when the mite population attains approximately two motile mites per leaf. This count is needed to establish the initial life stages distribution of the European red mite within the computer. Secondly, regular counts of the motile mite and *S. punctum* populations on a 7 to 10 day interval are needed to update the system. The mite population is sampled by determining the number of motile stages on at least 10 leaves collected around the tree and from five or more trees per monitored orchard. The number of trees selected to sample would depend largely on the size of the orchard. Croft *et al.* (1976) have described sample size estimates for the European red mite at varying densities and error margins. The trees selected for sampling the mite populations should be representative of the entire orchard in size and cultivar. If possible, cultivars which are sensitive to high mite populations such as 'Red Delicious' and 'York Imperial' should be selected because of the rapidity that European red mites increase on these cultivars. The *S. punctum* population is surveyed on the same trees as the mites and is accomplished by slowly walking around the periphery of the tree and recording the number of adults and larvae observed during three minutes. The survey for the *S. punctum* population should preferably be made before the leaves are collected for sampling the mite population because *S. punctum,* especially the adults, are very easily disturbed and will fall from the tree. Third, a record of daily maximum and minimum temperatures in the monitored orchard is needed to update the system periodically since the rate of development of the European red mite is primarily based on temperatures. The placement of a maximum-minimum thermometer or a hygrothermograph will suffice for providing such information. Finally, a record of any miticide or the mite suppressant dinocap used in the monitored orchard since the last count is determined.

Once this information is collected, it is sent or phoned to the nearest computer terminal that is tied in with the computer system at The Pennsylvania State University. Upon entering the data, the model is run and the output is returned to the terminal. The time required to enter the data, run the program, and return the resultant output for one orchard is about 10 minutes. The output (Fig. 6) that the grower or consultant receives contains the previous population estimates of the European red mite and *S. punctum;* the cumulative number of

Report for: Tyson Torway Block

Date of this run: 4-27-78

Date	Mites/Leaf All Stages	Mite-Days	Ladybird Beetles Per 3-min Count Adults	Larvae	A. fallacis Per 100 Lvs.
July 19	1.6	2.	4.	3.	2.
July 26	3.1	19.	6.	5.	2.
Aug 2	10.2	69.	11.	17.	3.
Aug 9	3.0	112.	12.	9.	12.
Predicted Population Trends					
Aug 10	2.6	114.	13.	9.	
Aug 11	1.4	116.	13.	10.	
Aug 12	1.4	117.	14.	10.	
Aug 13	0.4	117.	15.	11.	
Aug 14	2.9	120.	15.	11.	
Aug 15	1.4	122.	16.	12.	
Aug 16	4.0	126.	17.	12.	
Aug 17	4.3	130.	18.	13.	
Aug 18	3.7	134.	19.	14.	
Aug 19	2.5	136.	20.	14.	
Aug 20	1.2	137.	21.	15.	
Aug 21	0.0	137.	22.	16.	
Aug 22	0.0	137.	19.	14.	
Aug 23	0.1	137.	20.	15.	
Aug 24	0.1	138.	21.	16.	

The following recommendation, based on the predicted population trends, is provided only as a general guide. Other factors to consider are cultivar, amount of bronzing (if any), time of the season, tree vigor and soil characteristics.

RECOMMENDATION:

****

No miticide recommended. Continue regular counts of mites and Stethorus punctum (ladybird beetles). If the mite population increases to two or more adults per leaf, watch for an increase in the number of ladybird beetles.

****

Figure 6. Sample MITESIM computer output. The numbers of mites and *S. punctum* shown on Aug. 10 are the numbers obtained from the orchard count made on that date. Subsequent numbers are predictions. [Taken from: Mowery *et al.* (1977)].

motile mite-days, which is a measure of the amount of mite feeding inflicted upon a tree and is calculated by summing the mite density each day, the predicted sizes of both the European red mite and *S. punctum* populations for the next 14 days, and a management recommendation. The recommendation is intended only as a guide and is based on the size and rate of increase of the motile mite and *S. punctum* adult and larval populations, and the cumulative number of motile mite days. Other factors such as other predators (*e.g.* the predatory mite, *Amblyseius fallacis* (Garman), cultivar, time of the season, vigor of the tree, and crop load should be used in conjunction with this recommendation for deciding on what management tactic to employ.

VALIDATION OF MITESIM

After the construction of MITESIM was complete, the next step taken was to validate the model under actual orchard conditions. In 1976, the model was tested in three university-owned apple blocks. The European red mite and *S. punctum* populations were sampled on a 7 to 10 day interval starting in early June. Shown in Table 4 are the results obtained from one of the monitored blocks. MITESIM accurately predicted the trends in the mite populations throughout the months of July and August when the mite populations were increasing in substantial numbers. As shown in Table 4, the model recommended one half-spray application of a miticide which was applied on July 15 (28 g AI Cyhexatin 50WP per acre). Subsequent counts after the miticide spray revealed that the European red mite population would increase again. However, the model predicted that the size of the *S. punctum* population was large enough now to keep the mite populations from damaging levels, therefore, the model predicted that biological control by *S. punctum* was likely and recommended no further applications of a miticide. As shown in Fig. 7, MITESIM recommendations proved to be accurate. This apple block is a good example of how MITESIM can be used to determine the need for a miticide to supplement predation by *S. punctum* and not disrupt the balance of food for the natural enemies.

The use of MITESIM in Pennsylvania apple orchards cannot only result in reduced amounts of acaricides and increased savings to the grower, but also lends itself to less environmental pollution and hazard to the consumer. MITESIM is probably the prototype of other computer predictive systems that will be developed and implemented to aid growers, consultants, and orchard managers in decision-making regarding new pest management strategies and tactics for both insect and disease pests. With the aid of such systems to assist in the more

Table 4
Counts of motile European red mites
and *S. punctum* adults and larvae used to validate MITESIM.[e]

Date Sampled	Motile ERM/Leaf Actual	Predicted[a]	Recomm.[b]	*S. punctum* per min.
July 2	0.7			1.8
July 9	9.8	4.9	(2)	3.3
July 16[c]		12.4	(4)	
July 16[d]	3.2	1.7	(2)	22.0
July 26	7.1	10.9	(3)	38.0
Aug. 3	2.1	0.2	(1)	57.8
Aug. 11	1.1	0.0	(1)	27.8
Aug. 23	0.8	0.4	(1)	11.5

[a]Predicted on the date of the previous count.

[b]Recommendation (1): No miticide - continue regular counts;
(2): No miticide - watch orchard carefully;
(3): No miticide - biological control by *S. punctum* likely;
(4): Apply a miticide - biological control unlikely.

[c]Prediction made assuming no miticide.

[d]Prediction made assuming miticide on July 14. Miticide was actually applied on July 15.

[e]Taken from: Mowery *et al.* (1977).

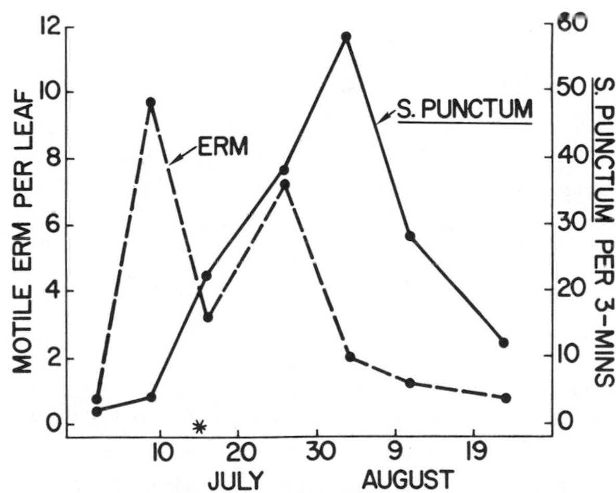

Figure 7. Average number of motile mites/leaf and *S. punctum* adults and larvae/3-min. count taken from an orchard in which MITESIM was validated. *Miticide July 15. [Taken from: Mowery *et al.* (1977)].

judicious application of pesticides, greater numbers of natural enemies, such as *S. punctum,* will move into apple orchards and further assist the grower in controlling the pests attacking his fruits.

Currently, much of the work on the integrated pest management in Pennsylvania is aimed towards indirect and induced pests such as plant feeding mites. Although much success has been achieved thus far in reducing overall pesticide usage through biological control of plant feeding mites, pesticide-prevailing programs are and probably will continue to be the prime tactic for maintaining the insect and disease pests attacking apples below economic injury levels. However, with the rapid advancement in technology coupled with the slow, but new development of more selective pesticides, a more efficient, economical, and ecologically sound integrated pest management system for Pennsylvania apple orchards which considers cost-benefits and environmental protection can be developed.

REFERENCES

Asquith, D. 1979. Systems approach and general accomplishments towards better insect control in stone and pome fruits. Cpt. 9, *In* "New Technology of Pest Control" (Ed. C. B. Huffaker). Wiley Intersci., New York. (In Press).

Asquith, D., and R. Colburn. 1971. Integrated pest management in Pennsylvania apple orchards. *Bull. Entomol. Soc. Am.* **17**: 89-91.

Asquith, D., and L. A. Hull. 1973. *Stethorus punctum* and pest-population responses to pesticide treatments on apple trees. *J. Econ. Entomol.* **66**: 1197-1203.

Asquith, D., L. A. Hull, and P. D. Mowery. 1975. The white apple leafhopper in Pennsylvania apple orchards. *Pa. Fruit News* **54**: 59-63.

Asquith, D., L. A. Hull, and J. W. Travis. 1976a. Apple, Test of Insecticides, 1973. *Insecticide and Acaricide Tests* **1**: 13-14.

Asquith, D., L. A. Hull, J. W. Travis, and P. D. Mowery. 1976b. Apple Insecticide Tests, 1974. *Insecticide and Acaricide Tests* **1**: 15-16.

Asquith, D., L. A. Hull, J. W. Travis, and P. D. Mowery. 1976c. Apple, Tests of Acaricides, 1974. *Insecticide and Acaricide Tests* **1**: 11-12.

Asquith, D., L. A. Hull, J. W. Travis, and P. D. Mowery. 1976d. Apple, Tests of Insecticides, 1975. *Insecticide and Acaricide Tests* **1**: 17-19.

Asquith, D., L. A. Hull, J. W. Travis, and P. D. Mowery. 1976e. Apple, Tests of Acaricides, 1975. *Insecticide and Acaricide Tests* **1**: 12-13.

Asquith, D., L. A. Hull, J. W. Travis, and P. D. Mowery. 1976f. Apple, Integrated Tests, 1975. *Insecticide and Acaricide Tests* **1**: 16-17.

Asquith, D., L. A. Hull, J. W. Travis, and P. D. Mowery. 1977a. Apple, Integrated Tests, 1976. *Insecticide and Acaricide Tests* **2**: 11-12.

Asquith, D., L. A. Hull, J. W. Travis, and P. D. Mowery. 1977b. Apple, Tests of Acaricides, 1976. *Insecticide and Acaricide Tests* **2**: 12-13.

Asquith, D., L. A. Hull, J. W. Travis, and P. D. Mowery. 1977c. Apple, Tests of Insecticides, 1976. *Insecticide and Acaricide Tests* **2**: 13-15.

Bode, W. M., D. Asquith, and J. P. Tette. 1973. Sex attractants and traps for tufted apple budmoth and redbanded leafroller males. *J. Econ. Entomol.* **66**: 1129-1130.

Cagle, L. R. 1946. Life history of the European red mite. *Va. Agric. Expt. Stn. Tech. Bull.* **98**. 19pp.

Colburn, R. B., and D. Asquith. 1971. Observations on the morphology and biology of the ladybird beetle *Stethorus punctum. Ann. Entomol. Soc. Amer.* **64**: 1217-1221.

Croft, B. A., S. M. Welch, and M. J. Dover. 1976. Dispersion statistics and sample size estimates for populations of the mite species *Panonychus ulmi* and *Amblyseius fallacis* on apple. *Environ. Entomol.* **5**: 227-234.

Hill, A., R. Carde, A. Comeau, W. Bode, and W. Roelofs. 1974. Sex pheromones of the tufted apple budmoth *(Platynota idaeusalis). Environ. Entomol.* **3**: 249-252.

Hill, A. S., R. T. Carde, W. M. Bode, and W. L. Roelofs. 1977. Sex pheromone components of the variegated leafroller moth, *Platynota flavedana. J. Chem. Ecol.* **3**: 369-376.

Horsburgh, R. L., and D. Asquith. 1968. Initial survey of arthropod predators of the European red mite in south-central Pennsylvania. *J. Econ. Entomol.* **61**. 1752-1754.

Hoyt, S. C. 1969. Integrated chemical control of insects and biological control of mites on apple in Washington. *J. Econ. Entomol.* **62**: 74-86.

Hull, L. A., D. Asquith, and P. D. Mowery. 1976. Distribution of *Stethorus punctum* in relation to densities of the European red mite. *Environ. Entomol.* **5**: 337-342.

Hull, L. A., D. Asquith, and P. D. Mowery. 1977a. The functional responses of *Stethorus punctum* to densities of the European red mite. *Environ. Entomol.* **6**: 85-90.

Hull, L. A., D. Asquith, and P. D. Mowery. 1977b. The mite searching ability of *Stethorus punctum* within an apple orchard. *Environ. Entomol.* **6**: 684-688.

Hull, L. A., D. Asquith, and P. D. Mowery. 1978. Integrated control of the European red mite with and without the mite suppressant dinocap. *J. Econ. Entomol.* (In press).

Lewis, F. H., and K. D. Hickey. 1967. Methods of using large airblast sprayers on apples. *Pa. Fruit News* **46**: 47-53.

Mowery, P. D., and D. Asquith. 1975. Computer simulation used to control orchard mites. *Sci. Agric.* **23**: 9.

Mowery, P. D., D. Asquith, and W. M. Bode. 1975. Computer simulation for predicting the number of *Stethorus punctum* needed to control the European red mite in Pennsylvania apple trees. *J. Econ. Entomol.* **68**: 250-254.

Mowery, P. D., D. Asquith, and L. A. Hull. 1977. MITESIM - A computer predictive system for European red mite management. *Pa. Fruit News* **56**: 64-67.

Pickett, A. D. 1959. "Utilizations of native parasites and predators". *J. Econ. Entomol.* **52**: 1103-1105.

Pickett, A. D., W. L. Putman, and E. J. LeRoux. 1956. "Progress in harmonizing biological and chemical control of orchard pests in eastern Canada". *Proc. 10th Int. Cong. Ent.* **3**: 169-174.

DEVELOPMENTS IN COMPUTER-BASED IPM EXTENSION DELIVERY AND BIOLOGICAL MONITORING SYSTEM DESIGN

B. A. Croft
Department of Entomology
Michigan State University
East Lansing, Michigan

S. M. Welch
Department of Entomology
Kansas State University
Manhattan, Kansas

D. J. Miller
Department of Entomology
Michigan State University
East Lansing, Michigan

M. L. Marino
Department of Plant Pathology
Michigan State University
East Lansing, Michigan

INTRODUCTION

Application of computer technology in agriculture and especially for integrated pest management (IPM) is increasing nationally in the United States. This is reflected by increased research to identify more efficient means of obtaining biological information on crops, pests and associated organisms (*e.g.* natural enemies of pests); integrating, interpreting and forecasting the possible meaning of these data for IPM and delivering this information rapidly to decision-makers in the field. Pest management simulation models are under construction and validation by research personnel on a nation-wide scale (Ruesink 1976, Tummala *et al.* 1976, Huffaker and Croft 1976). New extension scouting or biological monitoring systems for pests have been established in almost all agricultural states and on a variety of crops (Good 1977). Several of these monitoring programs have been coupled with computer-based, extension delivery systems on an experimental basis (*e.g.* Giese *et al.* 1975, Croft *et al.* 1976a). Implementation-research groups in many agricultural states have given a high priority to developing computer-based systems for IPM over the next 10-year period (*e.g.* Michigan, Bath *et al.* 1977; Indiana, Giese *et al.* 1975; New York, Barton 1977; California, Riedl and Allen 1978, Gutierrez 1978).

Although these groups and interested personnel in other states will develop implementation systems tailored to meet local needs, existing extension structures, specific cropping systems and funding constraints, there is much of common interest and philosophy in the design of these programs on a national scale.

In Michigan, development of a computer-based system of IPM including an environmental monitoring or weather network (Haynes *et al.* 1973), an extension delivery system (Croft *et al.* 1976a) and biological monitoring methods for a variety of organisms (Welch *et al.* 1978b, Olson *et al.* 1974, Bird *et al.* 1976) has been underway since 1970. Since these studies were reported, additional research on these IPM components has been completed. Discussion of certain aspects of these programs in this paper is presented in three sections: centralized delivery systems, hierarchically-distributed delivery systems and biological monitoring systems. Emphasis is on deciduous tree fruit research and extension projects, however this work encompasses several crop commodity areas (*e.g.* vegetables, small grains, field and forage crops; Bird *et al.* 1976) and involves a variety of discipline oriented groups besides entomology.

CENTRALIZED IPM SYSTEMS DEVELOPMENTS

At Michigan State University, several factors have contributed to the development of computer-based systems of IPM including (1) a common focus by system scientists, engineers and agricultural scientists on agroecosystem design and management, (2) interest in developing computer models of pest population dynamics, and (3) the need to process large amounts of IPM information gathered by statewide extension pilot projects for deciduous fruit (Olson *et al.* 1974) and field crop pests (Bird *et al.* 1976). This section outlines two major components of the resulting centralized system. The acronyms for these programs are PMEX and PETE.

PMEX

Using a cost-subsidized, interactive centralized computer system (based on a CDC 6500), an initial attempt at developing a computer-based extension delivery system was begun in 1974. It was first used in 1975 by an apple pest management project (Croft *et al.* 1976a) and since has been expanded to include a similar program on field crops (Brunner *et al.* 1978). PMEX (Pest Management Executive System) is a software program designed to provide users with access to pest management application and/or information programs and systems level documentation. It makes information storage and processing capabilities of computers available to users who have a limited understanding of these tools. Detailed computer control sequences and documentation are stored under keyword headings which can be easily used by untrained individuals (see Croft *et al.* 1976a and Brunner *et al.* 1978 for a more detailed discussion of PMEX development).

Remote site send-receive terminals and telephone playback devices, used in combination with the central processing program, constitute the communication networks of PMEX linking extension specialists, county agents, pest management personnel privately-employed fieldmen and growers (see structure in Croft *et al.* 1976a). As a result of PMEX, the temporal gap between research and application is appreciably reduced and there is an enhanced ability to deliver IPM information to decision-makers in the field.

There are four main types of facilities accessible via PMEX including communications features, biological monitoring summaries, environmental monitoring programs and predictive IPM models (Table 1). Communication features are extremely important components of any extension delivery system. They are especially valuable in the initial phase of developing such a system since access to more complex and detailed management models is usually limited. Using this system, memoranda and newsletter-type information may be disseminated more rapidly and to a wider spectrum of interested personnel than in a traditional delivery system. Biological information flows (Table 1) including data on population states of crops, pests and beneficial organisms can be transmitted quickly and efficiently to decision-makers in the field. Environmental monitoring programs aid in collecting and summarizing weather inputs received from a variety of U.S. weather network sources (*e.g.* NOAA, aviation weather service, Michigan agricultural weather network, etc.). These programs also are designed to be interfaced in a real-time mode or with a very short time delay to predictive IPM models for pest forecasting. IPM models (Table 1) are specific application programs designed to integrate biological monitoring, pest development, weather, crop status, economics and management data and to provide pest-managers with detailed information helpful in decision-making.

Since implementation of the PMEX system in 1975, use measurements have been collected and evaluated. Data from 1976 and 1977 are given in Table 2 which indicate that use by extension personnel averages *ca.* 2 times/week/user throughout the April 1 to October 1 growing season. Mean access time is from 12-15 minutes/session. The most commonly accessed programs (see Table 1) are FORECAST, CONTACT, INPUTER, DEGREEDAYS AND WTHRINPUT. These five programs accounted for more than 75% of the programs used in 1977. Although the relative proportion of use of these PMEX programs probably will remain high in the future due to their general applicability, it is expected that much greater use of more specific modeling, management and economic assessment programs will occur as these tools are more fully developed in the future.

Table 1.
Program types and examples of application programs available through the PMEX extension delivery system.

Communications	Biological Information	Environmental Features	Predictive Models
CONTACT: Sends messages of all types and in any format among PMEX users.	ALFALFA, SUGARBEET, ASPARAGUS, ONION, POTATO: provides pest summaries for these crops.	FORECAST: provides local daily weather forecasts, including four-day temperature predictions.	BLITECAST: a late blight forecasting model for timing of fungicide applications in potato production.
ALERTS: sends pest alerts and control strategy information to extension field staff.	BLACKLIGHT: summarizes insect information obtained through a blacklight trap program.	DEGREEDAYS: provides daily, weekly and seasonal degree day summaries from 58 sites.	WEEVILCOST: economic model of alfalfa production including variable cutting times, insecticide applications and management of alfalfa weevils.
AUTOMATIC SCHEDULING: allows users to receive crop pest information periodically without request.	BLITESUM: provides summaries of late blight program information generated by BLITECAST.	PRECIPSUM: provides daily, weekly and seasonal precipitation summaries from 58 sites.	MOTHMODEL: phenology model describing the development of the codling moth during a growing season.
GETMSG: allows user to obtain alerts and obtain messages through another user's PMEX account.	INPUTER: A generalized input program for biological data.	WTHRINPUT: inputs and displays raw weather data from 58 sites.	POTATOPEST: simulates potato losses caused by root-lesion nematodes and generalized insect defoliation.
MANUAL: delivers a complete set of PMEX instructions and program descriptions to a user.			MITEMODEL: prey/predator model for biological control of the European red mite by the predatory mite, *Amblyseius fallacis*.

Table 2.

PMEX system use patterns in 1976 and 1977.

Use measurement	Year	
	1976	1977
Total users	67	88
Total uses	4234	5398
Total uses/user	63.19	61.34
Real time use	861.1 hr.	1393.4 hr.
Use/user	12.85 hr.	15.83 hr.
Average session length	12.20 min.	15.49 min.
Total program	6412	8704
accesses/user	95.70	98.91
Help file	98	160
accesses/user	1.46	1.82
Most Frequently Accessed Programs		
	1976	1977
FORECAST	1971 [a]	2289
CONTACT	970	2030
INPUTER	655	738
DEGREEDAYS	509	467
WTHINPUT	439	655

[a] Represents total access/season.

Total developmental costs (including model development and adaptation costs) for PMEX to date have been *ca.* $86,000 and operating costs run about $25,000 per year (Edens and Klonsky 1977). Since much of the basic system development is completed, it is anticipated that developmental costs will decrease relative to operations costs in the future. Even for the 1975 versus the 1976 data, developmental costs declined by some 50%. Most future expenses will be due to new program development and general system improvements and maintenance. Telephone installation and equipment charges will decline once the system becomes more fully operational while the main portion of computer charges will be for operation in the future.

While PMEX has served as an excellent prototype extension IPM delivery system and has been heavily used and tested, it has limitations with respect to future needs and continued system development. In Table 3, a listing of strengths and limitations of this system are identified based on an economic and use assessment by personnel at Michigan State University (Edens and Klonsky 1977, Miller 1978).

Table 3.
PMEX strengths and limitations in relation to use, economic and operations assessment.

STRENGTHS	LIMITATIONS
2-way delivery of information with same communication channel useful in both directions	PMEX is a centralized system and as such has potentially reduced reliability and availability.
The system allows users to be contributors.	The prototype employed a single computer which was excessively large for many applications.
A minimum time delay between inquiry and response	Unrealistically high communication costs (exceed computation costs)
Users can interact with other users	The system is not capable of dealing with local needs in an economically efficient manner.
PMEX permits interrogation of computer-based data	The system does not represent an optimum combination of hardware and software.
The system permits judicious use of environmental information.	
Many biological and environmental events can be easily summarized.	
Information about management failures as well as successes is documented.	
Permanent machine-readable records of all observations and contributed data are readily available for use by research and extension staff.	
Not restricted to pest management—also available for crop and resource management.	
Users are not intimidated and make high use of PMEX programs.	
Field staff are the strongest supporters of continued development of a similar, more efficient system.	

In summary, PMEX does not represent a final product IPM delivery system but rather a 2-year prototype system which is part of an evolutionary program of systems development. PMEX features can be expected to improve in the future, certain ones will be deleted and others added as a less centralized, more distributed approach to delivery is sought (see later discussion). PMEX unit costs can be expected to decline as the number of users increase. Perhaps, more importantly, development and implementation of these systems in the future can be undertaken at a substantially reduced cost because of work already completed on PMEX.

It should be noted that PMEX is but one component of research and delivery in an agricultural management system although it is generalizable to many other areas of agricultural management (Croft *et al.* 1976a). Its true benefits can only be tested within the larger system. Accordingly, to better evaluate the net economic impact of such an IPM system, an analysis of the resources saved and crop losses averted should be obtained. Only when such a complete evaluation is completed with respect to the total use characteristics of the system, can the real benefits of such an approach to IPM delivery be determined.

PETE

As Table 1 shows, a number of predictive IPM models have been developed for use with the PMEX system. However, the extent of the pest complex on fruit, vegetables and (to a lesser degree) field crops means that much work is still necessary before comprehensive model-based management will be possible.

To facilitate the rapid development and extension implementation of phenological models a generalized Predictive Extension Timing Estimator (PETE) has been developed (Welch *et al.* 1978a). PETE allows research personnel to develop models with only a minimum of computer experience and effort. The researcher simply describes the organism by dividing it into key life stages (or other stages of convenience) and defining the duration of each stage in some (possibly temperature-dependent) units (*e.g.* degree days, calendar days, developmental units). Based on these data, critical timing "windows" for IPM are identified. Message texts to be sent to extension personnel in a predictive mode are entered. PETE then automatically updates each model using multi-point daily weather data acquired by the PMEX system. Any resultant pest alerts are automatically sent to appropriate extension field staff and researchers to aid in field decision-making.

There are presently 19 species contained in PETE, including the apple tree, apple scab disease, several arthropod pests of apples and two field crop pests. Incorporating models for new organisms is accomplished in four phases: (1) literature search, (2) parameterization, (3) validation and (4) implementation. Of the models initiated to date, several are in the developmental process, validation is underway for at least 10 species and several are being experimentally used in the field (Welch *et al.* 1978b).

HIERARCHICALLY-DISTRIBUTED SYSTEMS DEVELOPMENTS

Two major difficulties associated with the types of centralized systems discussed previously are high communications cost and low reliability/availability. The high communications cost is due to the necessity of paying data link charges (long distance telephone charges, dedicated line charges, etc.) for even the most trivial of uses of the system. In the PMEX system telephone charges were 50% greater than total computer charges. There are many types of operations which, although otherwise suitable, cannot be economically computerized on such a system. Examples include area or district extension office record keeping or accounting, scheduling of regional events, etc.

The entire system's dependence on the operation of the central processor results in low reliability and/or availability. There are many factors which interrupt or interfere with this operation. Some systems have several hours a day of scheduled down time; twenty-four hour staffing may not be economically feasible; time is needed for preventative maintenance, etc. Unscheduled down time may result from either hardware or software failures. While the former are generally quite rare, the latter are inevitable and often distressingly common for new systems. Another cause of reduced availability is heavy system loads during certain time periods.

All of these factors would not be problems if they were uniformly distributed in time. Unfortunately, experience has shown that the hours 8-9 a.m. and 3-5 p.m. are most critical to extension operations. Scheduled down time or startup difficulties are likely to interfere with the former; heavy use by students in mid-afternoon may affect the latter.

An alternative which solves these problems is to replace the centralized system with a hierarchical structure (*e.g.* Fig. 1). Such a system employs the small, cheap data processors recently created by digital engineers to distribute processing power to the regional, county and firm levels. These processors (along with certain pencil and paper methods) form a spectrum of data processing power reaching from low-cost, low-capability units to the high-cost, powerful computers previously discussed.

The key principle of the hierarchical system is this: if a number of people each have to solve some problems of low to moderate complexity, then a set of simple processors, working in parallel, can often do the job more efficiently than a single large processor working sequentially. If each individual has his own small

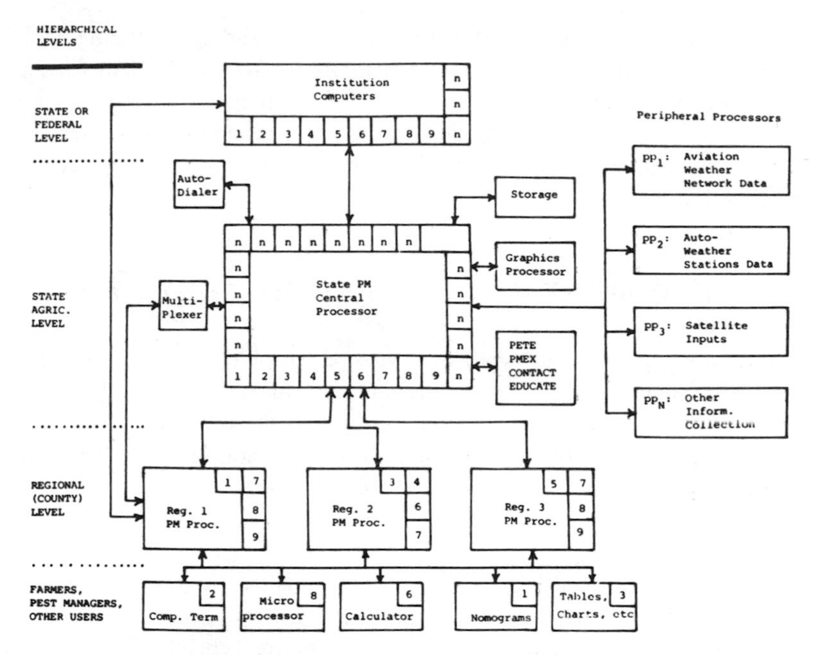

Figure 1. Hierarchical computer/processor based system for communication of IPM information. Numbers within processors illustrate programs residing within a given computer. These programs can be accessed by any other computer in the system via an automatic coupling mechanism [Adapted from: Bath *et al.* (1977)].

processor, he need not worry about the cost of communicating his problem to some distant site. Furthermore, if one of a group of such processors fails, the majority of users are not affected.

Within such a system institutional computers operate at state, multi-state or national levels, performing operations necessary to integrate IPM activities at that level (*e.g.* pesticide registration programs, environmental monitoring or national weather summaries). State IPM central processors manage a variety of statewide IPM functions including environmental network data from peripheral processors and executive systems features similar to those described in Table 1. Via telecommunication ties to regional processors, some subset of the program options at the statewide level are available at county or regional offices to meet specific needs in that area. At the lowest hierarchal level, firm level or implementation level hardware, software and IPM decision-making tools (computer terminals, microprocessors, calculators, nomograms, charts) assist pest managers working in the field. These field units can be updated, serviced and information gathered from them by regional processors. They may serve as input sources back to the state or higher level processors. This constitutes one important data gathering operation for interpreting and summarizing data from the field, in making appropriate recommendations from these data for regional IPM and in transmitting information back to fieldmen who must make the final decision for control and implementation of the program. These devices thus create closed information loops to aid in monitoring effective control or non-control in an on-line fashion.

HARDWARE IN THE HIERARCHICAL SYSTEM

A number of different types of devices exist which can be used at various levels of the hierarchy. Because of the rate of progress and diversification, it is difficult to form a strict classification system for these units; often there is no better guide than merely what the manufacturer "chooses" to call a device. In this section we will describe the more important of these units.

Minicomputers

These are full-fledged computers, albeit smaller than the systems which characterize many university computing laboratories. They are most readily identified by their small register size. Registers are typically 16 binary digits ("bits") wide although some machines are now reaching the 32 bit level. "Large" computers usually have registers in the 32 to 64 bit range. Memory sizes generally range from 16k to 256k words although there are exceptions at both ends. A word is an individually accessable unit of memory generally containing as many bits as a register; "k" is a common abbreviation for 2^{10} or 1024. The basic circuitry of minicomputers is extremely fast.

Minicomputers may be used for a variety of tasks in a hierarchical pest management system. Smaller ones ($15,000 - $30,000[1]) may be used for dedicated information management tasks. Examples might include maintaining files of weather data or biological survey data, transmitting memoranda from one part of the system to another, etc. The machine is "dedicated" in the sense that it performs one of these and nothing else. Larger minicomputers ($150,000 - $250,000) can coordinate the activities of a number of smaller minis, support time-sharing systems, and in general handle all but the most complex calculations or file storage tasks. These they can submit at their own initiative to a large processor. In short, a mini of this type could be the central controller for an entire state-level pest management delivery system.

---

[1] All prices in this section are approximate 1978 values.

Intelligent Terminals

These devices can be used in either a terminal mode or a stand-alone mode. In the former, they enable users to communicate with remote processors. Unlike most other terminals, however, these devices can edit, compute with, and otherwise modify the transmitted data. When not connected to a processor they can function as small computers in their own right. These machines usually have a high level language (often BASIC or APL) built into their hardware. This built-in characteristic distinguishes them from minis which rely on intermediate software translators to use these languages. These machines are slower than minis but their cost is also less ($6,000 - $12,000). Memory size is measured in terms of "bytes". A byte (eight bits) is sufficient to store one alphabetic character. Typical memory sizes are 4k to 65k bytes. A number of these terminals include cathode ray tubes (CRT) with graphics capabilities. There are also a variety of peripheral devices which can be connected to these units. These include printers, cassette or magnetic disk data storage units, x - y plotters, etc. Purchasing sufficient peripherals to create a usable system often doubles the complete price.

These systems are most suitable for regional office type applications. These might include accounting, record keeping, some types of statistical procedures, etc. For example, they could be used to collate reports from survey personnel within the regions, transmit a summary of that data to a central site (in terminal mode) and compose a list of counties which had similar problems previously or which must be notified now.

Intelligent terminals possessing graphic capabilities may be used to generate charts and graphs which present complex data concisely. Time savings in this activity alone may be 5 or 6 to 1.

Microcomputers

The major distinction between a microcomputer and other forms of computer is that the "intelligent" part (called the microprocessor) resides in a single large-scale integrated circuit as opposed to being made from many smaller circuits. This results in a system which is physically quite small. This makes microcomputers suitable for use as intelligent controllers of large devices. Indeed, microcomputers are often the source of the intelligence in the terminals we have just discussed.

Use of microcomputers as controllers has many potential applications at the firm level. Lillevik *et al.* (1977), for example, describes a field device for determining apple scab *(Venturia inaequalis)* severity using a microcomputer. Other uses might include calibration and control of chemical application equipment, monitoring of biotic and abiotic factors at the farm level, irrigation system control, etc.

Microcomputers are also being used as stand alone computers. Considerable interest has developed among hobbyists about what are called personal computers. These systems, which rival intelligent terminals in many respects, are available (often in easily assembled kit form) at prices ranging from $300 to $5,000. The latter figure includes peripherals (floppy disk storage, line printer, video system for limited graphics, etc.) and 16k to 32k bytes of extra memory. Personal computers are, in general, useful for tasks of moderate complexity like accounts payable, simple predictive models, etc. They are not suitable for supporting multiple terminals, large number "crunching" problems, or other tasks requiring high speed.

Programmable Calculators

There are two types of programmable calculators of interest to developers of information delivery systems. These are the larger desk top units and the small pocket calculators. At the higher end ($2,000 - $5,000) they overlap considerably in both price and performance with some of the microcomputer systems just discussed. Fisher and Welch (1974) review these calculators.

The small pocket programmables are potentially of great interest to system designers. These systems may be characterized by their cost ($100 - $500), program memory length (50-1,000 keystrokes), number of registers (10-100), methods of program entry (keyboard, magnetic tape reader, and read-only memories), and types of output (digital display and/or printers).

Programmable pocket calculators can be used directly in the field to solve problems requiring consideration of limited numbers of variables (*ca.* 5-10). Many important types of agricultural problems fall in this range: simplified economic threshold problems (*e.g.* Welch and Poston 1979), sampling methods (Johnson 1977), feed ration problems (Anon. 1977) evapotranspiration calculators (Kanemasu 1978), empirical pest control models (Croft 1975), etc. The application of pocket calculators to agriculture is reviewed by Welch and Poston (1979).

Other Equipment

Two other pieces of equipment facilitate communications within hierarchical systems: autodialers and multiplexers. Autodialers allow processors to place their own telephone calls. This will be of importance as agricultural management becomes more tightly controlled and key information must be disseminated to those in need in real-time, rather than relying on user requests as required with centralized systems which now exist (*e.g.* PMEX). Autodialers will enable processors to call one another when messages or information are to be transmitted; state level information can therefore be delivered to the regional level processor at any time of day or night to be readily and inexpensively accessed by regional staff and users. Multiplexers may be used to switch calls from one location to another. This allows users access to information available at a broader range of locations.

SOFTWARE IN THE HIERARCHICAL SYSTEM

The success or failure of any data processing system is dependent on its software. Like hardware, software in a hierarchical system is distributed throughout the management region. In Figure 1, numbers 1-9 associated with various processors represent nine example IPM programs (*e.g.* pest management models, economic assessment packages, etc.). Note that all programs do not reside in all processors.

When one processor is "down" information can be used from another computer holding that program. Such a fail-safe mechanism is necessary if this type of system is to gain widespread acceptability in agriculture. If a user dials the Region 1 IPM Processor and requests access to Program No. 1, this request will be handled from that processor. But if a request was made of that processor for Program 3, the call would be transferred to Region 2 and answered without the user ever knowing a transfer had been made. Commonly-used programs (such as numbers 7, 8 and 9) would reside in all or many processors. In sharing within the system, the user will be transferred up the hierarchical line to where the program resides. In multiple-state sharing systems, calls could be switched between states as the system grows and certain states took responsibility for development of specific aspects of a total pest management information network.

The location of software need not be static in the hierarchical system. Initially new programs would be implemented and debugged in a single processor. Then, as would be the case with any other type of information, the programs would be routed through the system to the points of use. The autodialing capabilities discussed previously would permit these transfers to be automated. If, over a period of time, demand for a particular program increased in some region, a copy of that program would be moved to that local processor. This would reduce the overall need for between-region switching and lower communications costs. Thus the system can respond dynamically to changing use patterns.

Another aspect of software is that facilities may be shared between levels. For example, simplified forms of decision models implemented on farm or county level processors may be backed up by more complex models situated higher in the hierarchy. Croft *et al.* (1976a) provides an example of such a backup system. Field research had established empirical relationships between initial prey-predator ratios and the probability of biological control among phytophagous mite populations in an orchard ecosystem. These results were summarized in the form of a static regression model which could be implemented in the form of a chart or simple calculator program. Although successful in 90% of the cases this model was limited in that it did not consider any population process such as development, consumption, dispersal, etc. This resulted in situations where the model was unable to reach a decision. To rectify this, a simulation model (Dover *et al.* 1979) was created. This model, residing in a central or regional processor could make recommendations in cases which the static model found ambiguous. Thus, the problem of orchard mite control would be shared between two levels in such a system.

This last example leads to another important point. Often the results or general patterns of behavior of complex models can be summarized in chart or nomogram or tabular form suitable for direct use by field personnel. Such material may be thought of as "software" for a human processor. Numerous examples of this exist in the literature. Shoemaker (1973) presented the output of an optimization model in tabular form; Carlson (1969) did the same for a Bayesian decision model. Valentine *et al.* (1976) used multiple simulation runs to generate probabilities tabulated in the form of a decision tree for gypsy moth control. Mowery *et al.* (1975) used a simulation model to produce a chart yielding spider mite control recommendations based on a two-step sampling procedure.

Cheap reproduction of such charts and their wide-spread use at the farm level can greatly reduce the processing burden at higher levels in the hierarchy. As with other software, it may be entirely self-contained or incorporate data acquired at or disseminated through higher levels of the system. It is through mechanism of this type of "software" that continuity is maintained with the vast body of extant extension publications.

BIOLOGICAL MONITORING

As Table 3 points out, information flows within delivery systems are bi-directional: raw biological data flow into the system concurrently with the outward flow of recommendations and summarized results. Programs for sensing or monitoring (Haynes *et al.* 1973) the states of a crop pest, and natural enemy populations must therefore be considered in the context of the associated delivery system.

This is the case particularly for monitoring systems involving high capital and/or operating costs. These costs must be balanced against the management improvements made possible for participating growers. Analyses of biomonitoring systems should take into account four general sets of factors: logistical, biological, statistical and economic. Logistical factors include all time delays involved in sampling, formulating, and delivering management recommendations. The biological set includes the developmental processes of the organisms in question (especially insofar as they generate a demand for monitoring), the distribution of target organisms, the biological interpretation of monitoring results and the biological effects of time delays. The term "statistical" refers to the stochastic aspects of sampling. The effects of spatial and sampling variation must be coupled with time delays so that the reliability of the monitoring data can be assessed. The above factors all interact in ways which affect the cost and revenues received by growers and by the monitoring agent. This necessitates an economic evaluation as well.

Because of the complexity with which these factors interact, models of monitoring and delivery systems can aid in predicting their behavior under a wide variety of operating conditions. By applying such analyses to alternative system designs, the optimal configuration of components can be selected. In this section we present such a model-based analysis of a mobile mite monitoring system for use in Michigan apple orchard IPM.

A MOBILE MITE MONITORING SYSTEM

Mobile monitoring systems may be thought of as a useful intermediate level in a hierarchical system. Although employed at the farm level for precise determination of local ecosystem states, their cost is shared on a regional basis. This permits concentration of equipment for local use which would be too costly for any one producer to acquire. Data obtained via a mobile system may be entered into a hierarchical system at either the farm or regional level. From there it may be summarized and distributed to other users of the system (researchers, extension personnel in adjacent areas, etc.). Operating management (demand projections, scheduling, accounting, etc.) for the mobile system may reside at the regional or state level.

In earlier publications the elements of a traditionally oriented IPM scouting system for deciduous fruit mite pests were described (Croft *et al.* 1976a; Olson *et al.* 1974). This traditional system is referred to here as Reduced System II. In this system a scout takes a leaf sample from the orchard, refrigerates it, and transports it to one of three centralized laboratories. These samples are then counted, control recommendations determined, and results mailed to the grower. Two alternatives (named the Full System and Reduced System I) were considered in addition to this scouting program. In the Full System a van equipped with a desktop programmable calculator, a computer terminal and laboratory equipment is dispatched to an orchard when the grower detects mite populations and requests a count. Two scouts take the sample, analyze it and make a recommendation using the data processing equipment. Count summaries are automatically transmitted to a central computer via the terminal at the end of the day. Reduced System I differs in that data is sent to the central computer at the end of the day vocally by telephone and entered by a technician.

MOBILE SYSTEM ANALYSIS SOFTWARE

The software used for this analysis consisted of two parts. The first part models the operation of the van in the field while the second analyzes the outputs of the simulation programs. Fig. 2 diagrams the mobile system and shows the activities carried out by each program.

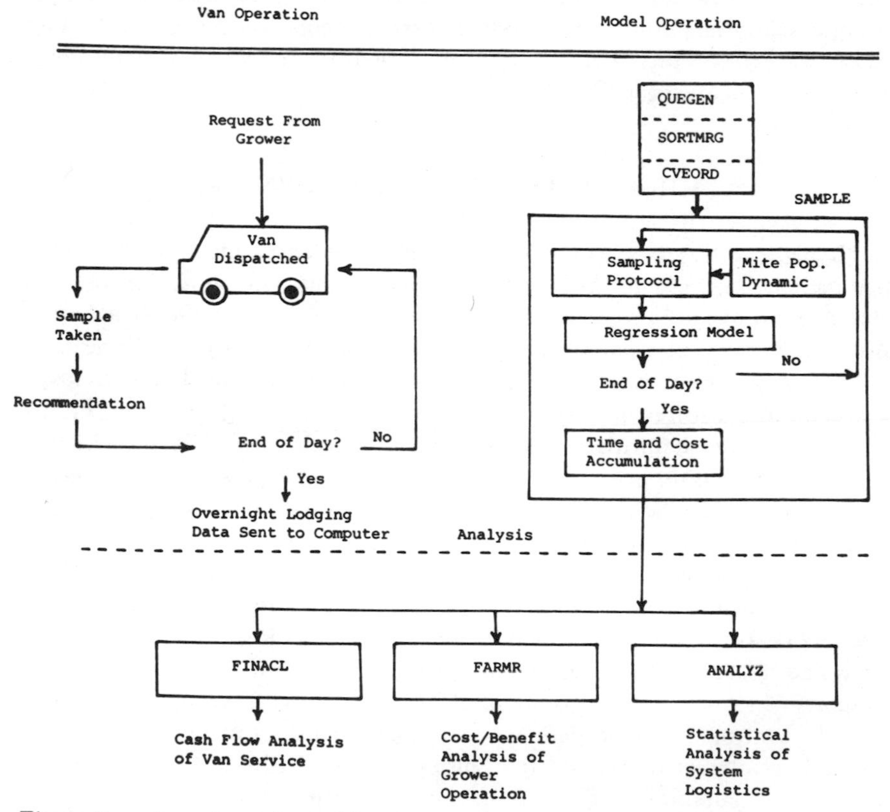

Figure 2. Operations of a mobile mite monitoring system including activities carried out by various simulation programs.

Program QUEGEN (Fig. 2) generates a series of requests for van service. These are based on a historical distribution of mite outbreaks broken down by date and area of state. The canned program SORTMERG then sorts the requests chronologically.

The next program, CVEORD, reorders each day's requests to minimize travel time. This simulates the scheduling activity which, as discussed above, might reside in a regional or state processor. The program uses matrices of travel times between different areas of the state based on the Michigan Department of State Highways and Transportation's road classification system and an accessibility study (Marino 1972).

The SAMPLE program simulates the operation of the van itself. Activities requiring fixed amounts of time (*e.g.* unpacking and repacking equipment, etc.) are simulated by adding fixed constants to an internal clock. Other activities (*e.g.* sequential sample collection and counting) are functions of mite density and, thus, of uncertain duration. Regression equations based on previous scouting experience (Croft 1974) and formulas developed by Welch (1977) are used to determine these times.

The densities of the predatory mite, *Amblyseius fallacis* (Garman), and the European red mite, *Panonychus ulmi* (Koch), at each site for the date of the sample request are created from cumulative probability functions for both mite types taken from records of mite densities. These densities represent the "true" state of nature. The waiting time between date of request and date sampled is then calculated. Predator and prey populations are updated for each day of the waiting period using a very simplified population model, based on that reported by Dover *et al.* (1979). The resulting densities are taken to be those determined in the sample.

Control recommendations are made using these densities and a static regression management model (Croft 1975). For prey-predator ratios where the regression model yields an indeterminate result, the "back-up" procedure discussed in the section on hierarchical system software is simulated. The model (1) assumes that detailed computer analysis would result in a no-spray recommendation and (2) increments all costs accordingly (including extra computer charges and time delays due to mailing recommendations). SAMPLE then calculates the date on which the recommendation would be implemented.

The probability of noncontrol (PNC) is next calculated for subsequent use in the cost analysis. This is determined from the formula:

$$PNC = \frac{1}{.95} \int_{CL} c(a, e, s, t)\ p(a, e \mid A, E)\ dade$$

where (A, E) is the estimated pair of mite densities (*A. fallacis* and European red mite, respectively) and (a, e) is the corresponding hypothetical true state of nature. The density function p gives the likelihood of the mite populations being in state (a, e) given that (A, E) has been measured. This distribution is assumed to be an uncorrelated bivariate normal with variances as determined by Croft *et al.* (1976b). CL is a 95 percent confidence region for this distribution. The function c equals one if populations in state (a, e) on the date sampled would exceed a nominal economic injury level (e = 15 mites/leaf) if control measure s is applied at time t; c is zero otherwise. Tabular values for c were determined by extensive computer studies.

As its final step, SAMPLE simulates the travel time delays involved in driving the van to the next sampling site. The total sampling process is then repeated. This continues until all requests for a given day are completed. The van is then sent to an overnight lodging by entering the travel time and noting the type (*i.e.* cost) of lodging used. Overnight lodging for the Full System and Reduced System I are campgrounds or motels. In Reduced System II, the van is sent to the nearest field station. Telephone, computer, and labor charges appropriate to each system's overnight data processing are calculated and totaled.

For Reduced System II the process is similar except that sample counting is simulated as occurring the day after sample collection. Appropriate extra delay times are included in the calculation of the date of recommendation, implementation, and the probability of non-control. Also accounted for are delays incurred by the mail delivery of recommendations.

All time, cost and density data generated by the simulation of each sample are combined into information "packets". These packets provide the input used by all the analysis programs.

The FINACL program analyzes van activities from the operator's viewpoint. This program is primarily an accounting system of costs and revenues. Its outputs include (1) a summary of simulation parameters (*e.g.* system used, number of growers served, activity time requirements, etc.) and (2) a cash flow analysis (Park 1973) containing data on revenues, costs depreciation, and rate of return on investment.

The program FARMR analyzes the costs to the grower. The mite counting service charge, the unit costs of spray material and application, and crop damage costs are entered into the program along with the packets. Tables similar to Table 4 are calculated on a per grower basis, per acre basis, and a statewide total basis. Other such tables are calculated under the assumption that the growers apply the recommendations on more that just the 10 acres actually monitored. Questionnaires have revealed that this is a common grower practice.

Table 4.

Cost analysis for growers.

1.	Cost of the strategy		
	A.	Cost of the control measures	
		1. Cost of material	----
		2. Application cost	----
	B.	Cost of the pest management system	
		1. Cost of making request	----
		2. Charge for monitoring	----
2.	Expected damage		----
		TOTAL COST TO GROWER	--------

The final program ANALYZ, tabulates logistical information about the sampling system. The program calculates a series of distributions based on the simulation parameters and the sample packets. These distributions include number of requests by area and date, number of requests per day, and amount of time spent counting, etc. Means, variances and percentiles are also printed.

SIMULATION RESULTS

The analysis programs generated a large amount of data. By examining graphs of this data (Fig. 3 and 4), several important trade-offs were investigated. Two of these are presented here:

Numbers of Growers Served

Michigan was divided into seven mite management areas and the number of growers served per area varied. The breakeven charge was defined as the per visit payment resulting in zero net profit after taxes to the van operator. Fig. 3 (Curve A) shows that his charge drops from $30.00 to $18.00 as the number of growers served per area increases from 30 to 60. These economics of scale can be passed on to the grower as a lower service charge.

However, this charge total is only a component of the cost faced by the grower. As the numbers serviced by a given system increase the time delay between date of request and date of monitoring also increases (Curve 3B). This can result in greater crop damage or control costs due to delay in applying control practices. Curve 3C shows the total per acreage costs increasing to a peak at 50 growers served even though the service charge is decreasing.

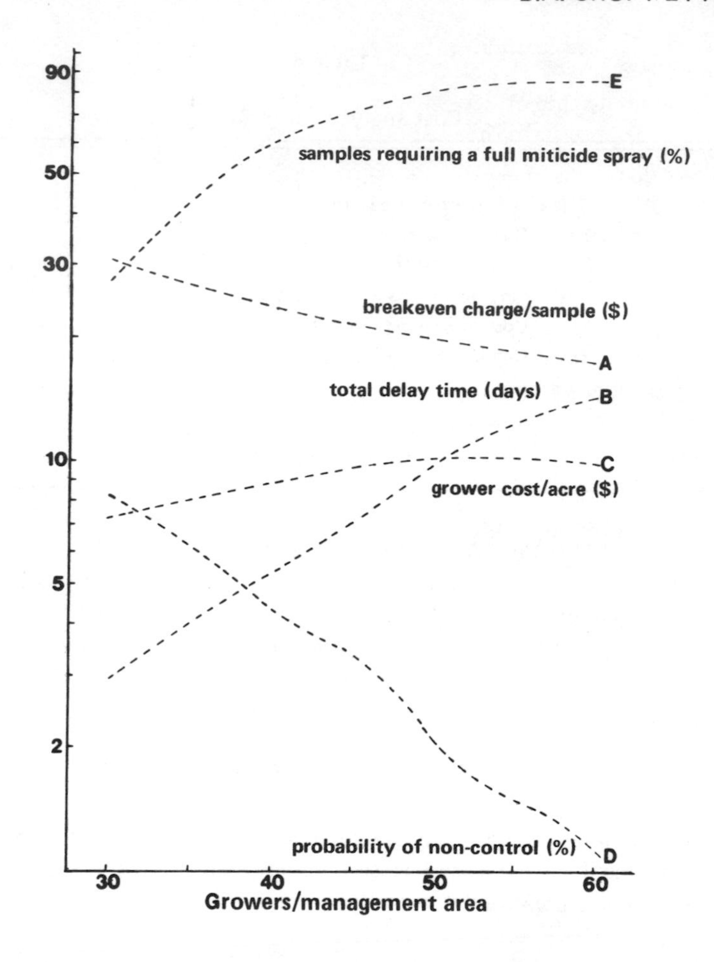

Figure 3. Output trade-offs involving the operation of a mobile mite monitoring system over a regional grower population.

The reduction in total cost above 50 growers per area seems to result from reduced service charges (Curve 3A) and reduced crop damages (Curve 3D) offsetting increased control costs (Curve 3E). At first Curve 3D may seem paradoxical; one would not intuitively associate better control with longer delays. The answer is that mite buildups which might have been controlled with low pesticide dosages earlier are instead receiving later, more expensive, but more certain, full dosages. This is the true significance of Curve 3E.

Sampling Accuracy

The immediate effects of increasing the accuracy of sampling are shown in Fig. 4. The abscissa is the width of the 80 percent confidence limits of the mean population density expressed as a fraction of the mean. As this decreases to 10, mean waiting time is little effected; below this, however, delay time is increased exponentially (see Curve 4A).

Curve 4B shows the effect on the service operator of increasing sample accuracy. Given a fixed service charge, profits decline rapidly as the width of the confidence limits is decreased below 10% of the mean. Curve 4C indicates that the total costs to growers increases rapidly if accuracy is increased beyond this point due to increased waiting time. An important conclusion which one can draw from these two graphs is that there are levels of accuracy which, far from being helpful to the grower can actually be detrimental.

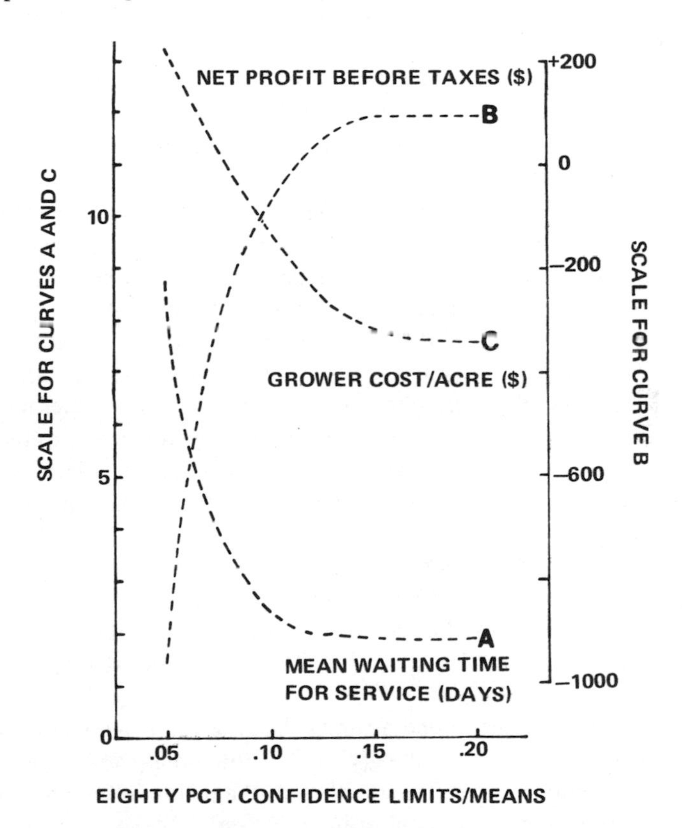

Figure 4. Effects on sampling accuracy relative to net profit, grower costs and mean waiting time for a mobile mite monitoring unit.

A SIMPLIFIED APPROACH

In the previous paragraphs a complex model was presented which estimated the performance of a monitoring system. This model required a great investment in time and effort to develop although it permits system behavior to be analyzed in great detail. To compliment this approach, Welch (1977) has developed a simpler procedure which allows a preliminary screening of design alternatives with reduced effort. The procedure also closely ties the design of biomonitoring programs to that of the overall extension delivery system.

The method is based on the concept of a monitoring unit. This is a geographical unit within which sampling is carried out. The sampling results are assumed to apply uniformly within the monitoring unit. The size of the unit determines the spatial resolution of the entire biomonitoring programs. For species subject to localized outbreaks this resolution must be high; monitoring units would be farm-sized or smaller. For other organisms, monitoring units might be of regional scale. Thus monitoring systems for varied pest complexes would have a structure congruent with that presented previously for hierarchical extension delivery systems. This similarity could be expected to greatly facilitate the flow and processing of information.

Factors accounted for in the simplified procedure include: (1) apparent variability among monitoring units, (2) economic risks, (3) management system time delays, and (4) sampling effort. Variability among monitoring units results from various biological causes (*e.g.* micro- and meso-habitat variation and climatic differences, etc.) as well as the stochastic attributes of the sampling procedure. Economic risk to the grower is a direct result of the uncertainty this variation engenders. Several examples of time delays were presented in the previous section. Sampling effort or intensity refers to the inputs to the sampling process. It would, therefore, exclude components like accuracy which is an output characteristic. It would, however, measure the effort required to achieve any given level of accuracy.

Needless to say, there are many different measures which might be used to quantify the four types of factors just discussed. Table 5 lists some of these. In the following discussion we shall assume that one measure of each type has been selected for consideration. The indices will be abbreviated v, i, t, r as shown in Table 5.

The analysis to be presented assumes that the following relations hold for the two functions $t = h(r, v)$ and $i=g(v, t)$. The functions g and h are assumed to be defined, continuous, and differentiable over the region of interest.

The function g relates monitoring unit variability and time delays to intensity of monitoring. Variability and delays are closely related to the

Table 5.
Some representative variables from each of the four important design parameter classes.

Variability within monitoring units (v)
Variance of monitored variables
Geographic size of monitoring unit [a]
Dissimilarity index of sites within unit
Etc.
Sampling Effort (i)
Man-hours spent monitoring
Number of samples taken per unit time
Total monitoring expenditures per season
Etc.
Management system time delays (t)
Average time from monitoring to action
Maximum time from monitoring to action [a]
Etc.
Economic risk (r)
Expected total loss
Probability of exceeding economic injury level [a]
Probability of sustaining focus of loss
Etc.

[a] These indices used in example.

reliability of the information ultimately delivered to the grower; increases in either one can degrade usefulness of the results. Beyond selecting homogenous monitoring units, reductions in v can only be achieved by increasing sampling effort. This is stated in inequality 1. Trying to improve reliability by decreasing time delays also requires an increased effort as shown in inequality 2.

$$(1)\quad \frac{\partial i}{\partial v} < 0 \qquad\qquad (2)\quad \frac{\partial i}{\partial t} < 0$$

$$(3)\quad \frac{\partial t}{\partial r} > 0 \qquad\qquad (4)\quad \frac{\partial t}{\partial v} < 0$$

The function h yields the maximum allowable time delay in terms of monitoring unit variability and permissible risk level. It is determined primarily by the biology of the organism. As delays increase, the more conceivable the possibility (*i.e.* risk) that the system has entered a damaging state. Inequality 3 simply states that the higher the permissible risk, the longer the permissible delay. Inequality 4 maintains, on the other hand, that the greater the uncertainty due to monitoring unit variation, the quicker we must act to avoid high levels of risk.

Welch (1977) shows mathematically that when these inequalities hold, it is possible to solve numerically for any pair of variables given any other pair (with one minor exception). By way of example, he constructs such a solution for the mite monitoring problem. Three of the indices chosen are indicated in Table 5. A relative index was used for sampling intensity; it is proportional to the frequency of sampling and the density of sampling points. The solution was put in graphical form by plotting contours of permissible delay and sampling intensity on a grid of monitoring unit size and economic risk (Fig. 5).

The resultant nomogram has a variety of uses. For example, for a given maximum level of permissible risk, one can determine at a glance the allowable tradeoffs between delay and monitoring unit variability (represented here by size). Or, given a system whose time delays and monitoring unit sizes are known, one can estimate its associated risk and sampling intensity. This was done for Reduced System II (Fig. 5) based on the average distance traveled between

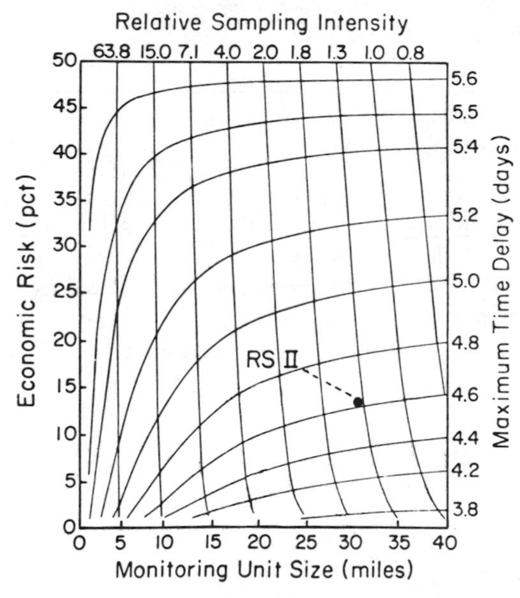

Figure 5. Graphical form relationships between 4 indices relative to sample design for a mobile mite monitoring unit.

samples as determined by ANALYZ. The estimated risk level (12%) agreed quite well with the 11% value obtained by examining the frequency with which actual mite counts exceeded a nominal economic threshold (Croft and McGroarty 1978). Other uses of these nomograms are elaborated in Welch (1977).

In a sense, the complementary nature of the simulation versus nomogram approach parallels that often cited for simulation versus optimization in other modeling studies. Indeed, the nomogram can be used as an optimization tool. By plotting constraints (such as maximum risk etc.) feasible regions can be identified. The less-detailed nature of the mathematical formulations makes the use of formal optimization methods possible. Given this dual approach, the monitoring system designer can move rapidly to consideration of the limited set of alternatives most likely to meet his needs.

SUMMARY

Because of the need to rapidly acquire, process, and disseminate large amounts of IPM data the trend toward increasing computer use can be expected to continue. System configurations will reflect statewide or regional operating conditions, expertise, and economics. Given, however, the advantages of reliability and flexibility inherent in hierarchical systems, it seems probable that this form of organization will become widespread.

As such systems develop, they will have impact on the design of biological monitoring systems. Along with abiotic monitoring, biomonitoring is the source of much of the data flowing through these IPM systems. Features such as biomonitoring, economics, logistics, and statistical properties will receive increasingly detailed attention. We feel that the simulation and graphical techniques presented herein will have important roles in these developments.

ACKNOWLEDGMENTS

This research was supported in part by NSF-EPA grant BMA-04223 to the University of California and Michigan State University. The findings reported herein are the opinions of the authors and not necessarily those of the University of California, Michigan State University, NSF, or EPA.

REFERENCES

Anonymous. 1977. "Programmable calculator becomes farm tool". *The Iowa Stater,* November, Pg. 5, Iowa St. Univ., Ames, IA.

Barton, D. W. 1977. When we get the computer educated . . . *NY Food and Life Science Quote* **10**:3, 20.

Bath, J. E., G. W. Bird, B. A. Croft, D. L. Haynes, S. H. Gage, K. Dimoff, R. F. Ruppel, and R. L. Tummala. 1977. Development of a processor based hierarchical system for communication of integrated pest management information. EPA grant. Mich. State Univ. 119 pp.

Bird, G. W., M. J. Dover, and M. Saratte. 1976. On-line pest management project annual report. *Mich. State Univ. Tech. Rept.* 3. 119 pp.

Brunner, J. F., B. A. Croft, S. M. Welch, M. J. Dover, and G. W. Bird. 1978. A computer based extension delivery system: Progress and outlook. *EPPO Plant Protect. Bull.* (In press).

Carlson, G. A. 1969. A decision theoretic approach to crop disease prediction and control. Ph.D. Dissertation, University of California, Davis.

Croft, B. A. 1974. Unpublished data. Mich. State Univ., East Lansing, Mich.

Croft, B. A. 1975. Tree fruit pest management, Cpt. 13, pp. 471-507 *In* "Introduction to Insect Pest Management." (Eds. R. L. Metcalf and W. Luckemann). Wiley Intersci., New York. 587 pp.

Croft, B. A., and D. L. McGroarty. 1978. The role of *Amblyseius fallacis* in Michigan apple orchards. *Mich. Agric. Expt. Sta. Rept.* 333. 24 pp.

Croft, B. A., J. L. Howes, and S. M. Welch. 1976a. A computer-based, extension pest management delivery system. *Environ. Entomol.* **5**:20-34.

Croft, B. A., S. M. Welch, and M. J. Dover. 1976b. Dispersion statistics and sample size estimates for populations of the mite species *Panonychus ulmi* (Koch) and *Amblyseius fallacis* (Garman) on apple. *Environ. Entomol.* **5**: 227-234.

Dover, M. J., B. A. Croft, S. M. Welch, and R. L. Tummala. 1979. Biological control of *Panonychus ulmi* (Koch) (Acarina: Tetranychidae) by *Amblyseius fallacis* (Garman) (Acarina: Phytoseiidae) on apples: a prey/predator model. *Environ. Entomol.* (In press).

Edens, T. C., and K. Klonsky. 1977. The pest management executive system (PMEX): Description and economic summary. *Mich. State Univ. Agr. Exp. Sta. Rept.* 96 pp.

Fisher, P. D., and S. M. Welch. 1974. When to use a programmable calculator or a mini-computer to automate a system. *Electronics* **47**:100-103.

Giese, R., R. M. Peart, and R. M. Huber. 1975. Pest management. *Sci.* **187**: 1045-1052.

Good, J. M. 1977. Integrated pest management - a look to the future. *USDA Bull. ESC* 583. 18 pp.

Gutierrez, A. P. 1978. Applying systems analysis to integrated control. *Calif. Agric.* **32(2)**:11.

Haynes, D. L., R. K. Brandenburg, and P. D. Fisher. 1973. Environmental monitoring network for pest management systems. *Environ. Entomol.* **2**:889-899.

Huffaker, C. B., and B. A. Croft. 1976. Integrated pest management in the U.S.: Progress and promise. *Environ. Hlth. Prosp.* **14**:167-183.

Johnson, D. C. 1977. Sex-ratio estimation, sequential sampling and programmable pocket calculators. *Bull. Entomol. Soc. Am.* **23**:251-254.

Kenemasu, E. T. 1978. Personal communication to S. M. Welch.

Lillevik, S. L., A. L. Jones, and P. D. Fisher. 1977. A predictive field instrument for agricultural production. *Microcomputer Conference Record. IEEE Catalog No.* 77CH1185-8, pp. 137-142.

Marino, M. L. 1972. An accessibility study of the northwest region of Michigan's lower peninsula. *Mich. State Dept. Highways and Transportation Rept.* 87pp.

Miller, D. J. 1978. Unpublished data. Mich. St. Univ., East Lansing, Mich.

Mowery, P. D., D. Asquith, and W. M. Bode. 1975. Computer simulation for predicting the number of *Stethorus punctum* needed to control the European red mite in Pennsylvania apple trees. *J. Econ. Entomol.* **68**:250-254.

Olsen, L. G., C. F. Stephens, J. E. Nugent, and T. B. Sutton. 1974. Michigan apple pest management annual report: 1973. *Mich. Exten. Serv. Rept.* 57pp.

Park, W. R. 1973. Cost Engineering Analysis. John Wiley and Sons, Inc., N.Y.

Riedl, W. H., and W. W. Allen. 1978. Implementing integrated pest management in California. *Calif. Agric.* **32(2)**:8-9.

Ruesink, W. G. 1976. Status of the systems approach to pest management. *Ann. Rev. Entomol.* **21**:27-44.

Shoemaker, C. 1973. Optimization of agricultural pest management III: Results and extension of a model. *Mathematical Business* **18**:1-22.

Tummala, R. L., D. L. Haynes, and B. A. Croft. 1976. Eds., "Modeling for Pest Management: Concepts, Applications and Techniques." *Mich. State Univ.* 247 pp.

Valentine, H. T., C. M. Newton, and R. L. Talerico. 1976. Compatible systems and decision models for pest management. *Environ. Entomol.* **5**:891-900.

Welch, S. M. 1977. The design of biological monitoring systems for pest management. Ph.D. Thesis, Michigan State University, East Lansing.

Welch, S. M., B. A. Croft, J. F. Brunner, and M. F. Michels. 1978a. PETE: An extension phenology modeling system for management of a mite-species pest complex. *Environ. Entomol.* **7**:487-494.

Welch, S. M., B. A. Croft, J. F. Brunner, M. J. Dover, and A. L. Jones. 1978b. The design of biological monitoring systems for pest management. *EPPO Plant Protect. Bull.* (In press).

Welch, S. M., and F. L. Poston. 1979. The programmable calculator as a tool in farm-level pest management. *Bull. Entomol. Soc. Am.* (In press).

INDEX